Chemical Stabilisation of Soft Soils

Chemical treatment is often used in ground engineering for the rapid stabilisation of soft soil. Although different chemicals can be used, there are significant commonalities in their underlying principles and reactions. This book provides a holistic perspective on chemical stabilisation, covering its history, chemistry, factors affecting quality in the mixing processes, the properties and constitutive behaviour of improved soils, quality control, and design and construction.

Uniquely, this book draws out the interrelationships between the various methods of chemical stabilisation of soils. A key example is the relation between the water content and strength, constitutive behaviour, and uniformity of mixing. Operating parameters may produce conflicting results. For instance, a high water content promotes flowability and thus miscibility, but can also result in binder–soil segregation due to density differences. Furthermore, higher water content tends to degrade the strength of the treated soil. Hence, it is important to find an optimum balance.

This book is essential particularly for geotechnical engineers who are involved in the design and execution of chemical improvement of soil, as well as researchers.

Fook-Hou Lee is a full professor at the National University of Singapore as well as a professional engineer with extensive consultancy and advisory experience involving chemical stabilisation for deep excavations in soft soil.

Chemical Stabilisation of Soft Soils

Theory and Practice

Fook-Hou Lee

CRC Press
Taylor & Francis Group
Boca Raton London New York

CRC Press is an imprint of the
Taylor & Francis Group, an **informa** business

Designed cover image: Fook-Hou Lee

First edition published 2026
by CRC Press
2385 NW Executive Center Drive, Suite 320, Boca Raton FL 33431

and by CRC Press
4 Park Square, Milton Park, Abingdon, Oxon, OX14 4RN

CRC Press is an imprint of Taylor & Francis Group, LLC

ISBN: 978-1-032-25964-2 (hbk)
ISBN: 978-1-032-25965-9 (pbk)
ISBN: 978-1-003-28588-5 (ebk)

DOI: 10.1201/9781003285885

Typeset in Times
by Apex Covantage, LLC

Dedicated to my late wife, Linda Lim, without whose unwavering support and encouragement, this book could not have been completed.

Contents

Acknowledgements

I would like to express my deepest gratitude to my family for their unwavering support, encouragement, and motivation throughout this journey. I am particularly indebted to my late wife, Linda, who sacrificed many hours of quality time with her husband so that he can concentrate on writing this book.

I would also like to express my appreciation to the many people who have assisted me during the writing of this book. In particular, I would like to acknowledge the contribution of Jonathan Daramalinggam, Chang Kaan Loh, and Victor Chee Wee Ong in sourcing for appropriate photographs and Figures.

Finally, to my colleagues at Taylor & Francis, especially Tony Moore, I would like to express my appreciation for their patience and forbearance.

1 Introduction

1.1 HISTORICAL BACKGROUND: IN AND EX SITU MIXING

The idea of using chemicals to improve and harden soil has been known for a few thousand years. Lime mortar mixed with sand and clay were discovered on the floors of houses at the archaeological site of Lepenski Vir in Serbia, dating back to 7500 BC (Radić et al. 2008). Lime was also used in the construction of the pyramids at Shersi in Tibet (Miller 2000), as well as in Egypt, China, Greece, Rome and other ancient civilisations. In many of these construction-related applications, lime was commonly used in the admixture with soil in order to bring about pozzolanic reactions, which led to the production of cementitious products, that comprise mainly calcium silicate hydrate and calcium aluminate hydrate.

In most of these historical applications, the admixing of soil and lime was probably conducted ex situ, that is, out of the ground. In more recent times, lime admixture has also been used to stabilise high plasticity soils, which are susceptible to large volumetric changes, as well as to improve internal cohesion of low plasticity clays, which are susceptible to dispersion and erosion. An example of such an application is the stabilisation of the banks of the Friant-Kern Canal (e.g. Gutschick 1978; Howard and Bara 1976) (Figure 1.1). The banks of the canal were found to be susceptible to cracking and sliding, owing to the presence of montmorillonite in the soil, which led to repeated swelling and shrinkage. The solution consisted of treating the soil with lime to reduce the plasticity of the soil, thereby eliminating the large volumetric changes of the soil. The roads and bottom of the canal were treated in situ while the banks were treated by first excavating the soil, mixing it with lime and then re-compacting the lime-admixed soil in layers. According to Howard and Bara (1976)

> all the material that was to be stabilized with lime and recompacted was removed by a benching operation. A series of long sloping benches or ramps were cut from the top of the bank down to the canal bottom with the cut extending far enough into the slope to remove the entire depth of required excavation material. Two percent quicklime was spread over the bench surface and 0.3 m (1 ft) of material from the bench was mixed with the quicklime, and the lime-clay mixture was pushed into the canal bottom where the remaining oversize material was removed.

Ex situ mixing was also used in the improvement of the Mississippi River levees (e.g. Fleming et al. 1992; Gutschick 1978) and in Arkansas and Illinois, where repeated slides had occurred due to the presence of expansive clays. The treatment consisted of excavating down to a level just below the slip plane, mixing the soil with lime offsite and then re-placing and compacting the admixed soil. In these

DOI: 10.1201/9781003285885-1

works, the lime content used typically ranges from 2% to 5%, the purpose being to reduce plasticity index, expansion and shrinkage of the soil as well as dispersion and tendency to erode, rather than strength gain. Nonetheless, Howard and Bara (1976) reported substantial strength gain from undisturbed soil. Whereas the untreated soil had an unconfined compressive strength of 110kPa, the treated soil had strengths ranging from 870kPa to 3560kPa, despite the relatively low lime content.

Although ex situ mixing methods are still used, admixing is increasingly done in situ, that is without removal of the soil to be treated, in order to avoid additional earthwork involved in removing soil from the ground and then re-placing the admixed soil. By obviating the need for a mixing site, encroachment into surrounding areas is also minimised. In situ mixing operations can be divided into deep and shallow mixing, depending upon the depth of the soil layer to be stabilised.

The origin of modern in situ soil mixing method is generally attributed to a patent on mixed-in-place pile assigned to Intrusion-Prepakt Inc. (Liver 1962). As Figure 1.2 shows, the design featured a rotatable shaft with a drill or auger bit at the bottom and cutting/mixing blades along the stem. The equipment is jacked into the ground while the shaft is being rotated, thereby cutting and loosening the soil in a cylindrical cavity, down to a desired depth. Chemical grout is introduced into the ground via an opening at the drill bit. This remains the basic concept of the deep mixing, which is still widely today.

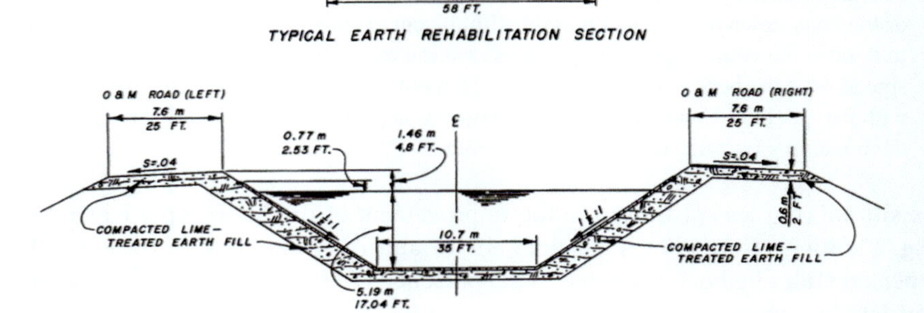

FIGURE 1.1 Cross-section of Friant-Kern Canal showing lime-treated zones along the bottom, slopes and service roads.

(*Source:* Howard and Bara 1976).

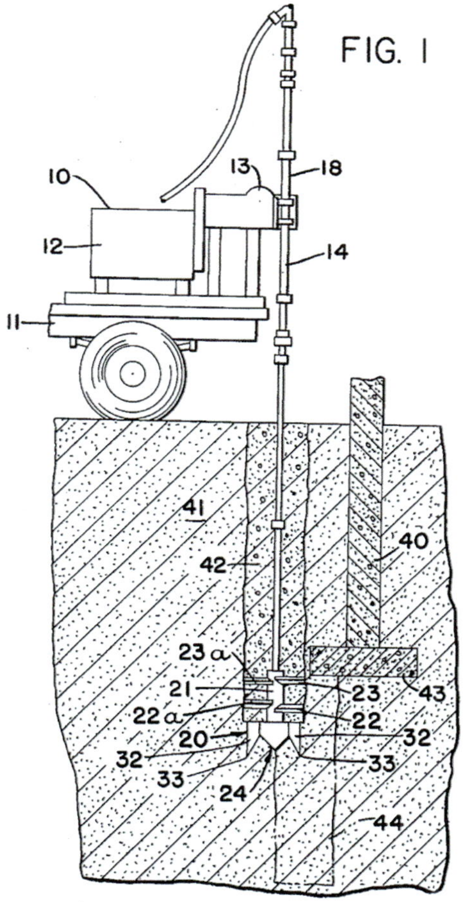

FIGURE 1.2 Mixed-in-place pile equipment, showing drilling bit or auger at the tip and cutting-mixing blades on the stem.

(*Source:* Liver 1962).

1.2 DEEP/SHALLOW MIXING AND JET GROUTING

In BS EN 14679 (2005), deep soil mixing (DSM) is considered to be admixture treatment of soil at a depth of at least 3m below ground surface. Treatment for depths of less than 3m below the ground surface is considered to be shallow soil mixing (SSM). One type of equipment for SSM followed the concept described by Liver (1962), but with a short rotary shaft and large auger, the latter with diameter up to 3.7m (Topolnicki 2004). Other concepts such as bucket mixers and mixing/cutting wheels, both of which differ significantly from the vertical auger-shaft concept, have also been used for shallow mixing equipment (e.g. Raito 2018).

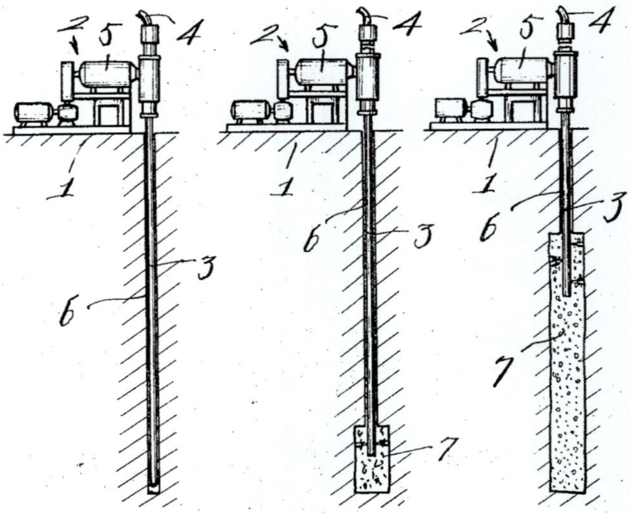

FIGURE 1.3 Schematic of Nakanishi's (1974) jet-grouting method, showing (a) the drilling of a pilot cavity, (b) withdrawal of the jet-grouting stem, cutting of the jet-grouting cavity, and (c) advanced stage of withdrawal with the jet-grout column largely formed.

(*Source:* Nakanishi 1974).

Another concept which is widely used in deep chemical stabilisation is jet grouting, which is widely considered as a competing technology to DSM. The early development of jet grouting is widely attributed to Miki (1973) and Nakanishi (1974), who termed it as a "chemical churning pile" method. In this method, the soil is cut and mixed with the grout by means of high-pressure grout or water jets, instead of cutting blades (Figure 1.3). Although part of the cut soil and water is expelled from the cut cavity in the form of slurry, Nakanishi's (1974) patent states that the process involves forming "*a plurality of successive columns . . . by pumping an earth solidification and densification liquid agent into the earth and soil formation so as to mix the liquid agent with the earth and soil,*" which indicates that some degree of mixing was intended.

The previous discussion only gives a broad classification to the available soil mixing methods. In reality, many variants exist as proprietary ground treatment technologies. For instance, the Jet Mechanical Mixing (JMM) or RAS-JET system (Raito 2007) combines the concepts of deep mixing and jet grouting by incorporating jet-grouting nozzles at the tips of the mixing blades. These extend the cutting and mixing reach, thus allowing treated soil columns of larger diameter to be formed, as well as allowing more grout to reach the periphery of the columns.

1.3 AREAS OF APPLICATION

Chemical stabilisation is used in a large variety of geotechnical engineering applications. Depending upon the applications, the treatment is often directed towards achieving the following objectives:

1 Increasing the strength and stiffness of the soil, in order to prevent failure and minimise ground movement in foundations, embankments, slopes and excavations.
2 Reducing permeability so as to minimise groundwater intrusion.
3 Improving stability of unstable, dispersive or expansive soils.
4 Minimising liquefaction potential under earthquake conditions.

1.3.1 Foundation Improvement

Chemical stabilisation is widely used in the strengthening of foundations for structures of large areal extent, such as embankments, caissons and storage tanks, as a competing technology to stone columns. Topolnicki (2016) reported the usage of DSM to improve the foundation soils beneath approach embankments for highways. Bergado et al. (1999) also reported the use of DSM for improvement of soft clay underlying the embankment for the Bangna-Bangpakong Highway in Thailand. Instances of DSM usage for embankment foundations have also been reported in Kitazume and Terashi (2013); these include river embankments in Chiba, Saitama and Ibaraki Prefectures as well as road embankments in Chiba. Although DSM was used in most of the cases, Kitazume and Terashi (2013) reported an instance involving the use of jet grouting for a river embankment in Tokyo. In many of these applications (e.g. Bergado et al. 1999; Topolnicki 2016), the DSM columns were installed as closely spaced but not contiguous columns (e.g. see Figure 1.4). On the other hand, in many of the instances reported by Kitazume and Terashi (2013), the columns were installed in contiguous grid or block assemblies (Figure 1.5). The differences in the design are attributable

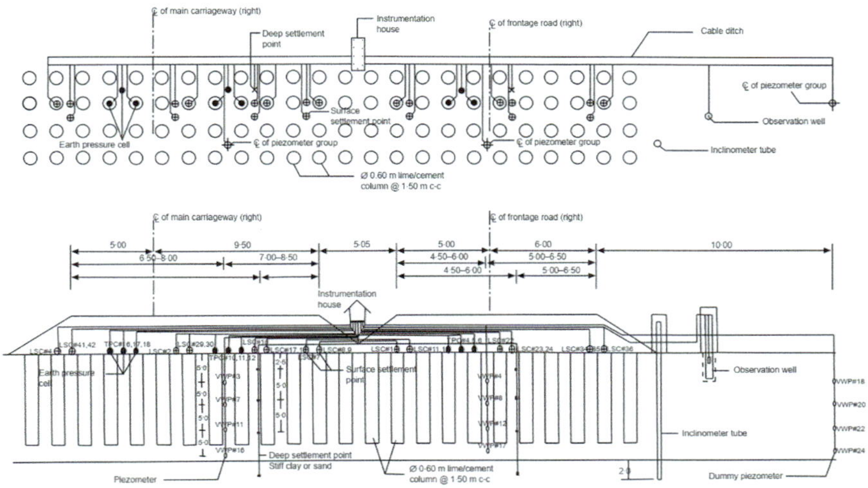

FIGURE 1.4 Layout of DSM columns for the embankment of the Bangna-Bangpakong Highway in Thailand.

(*Source:* Bergado et al. 1999).

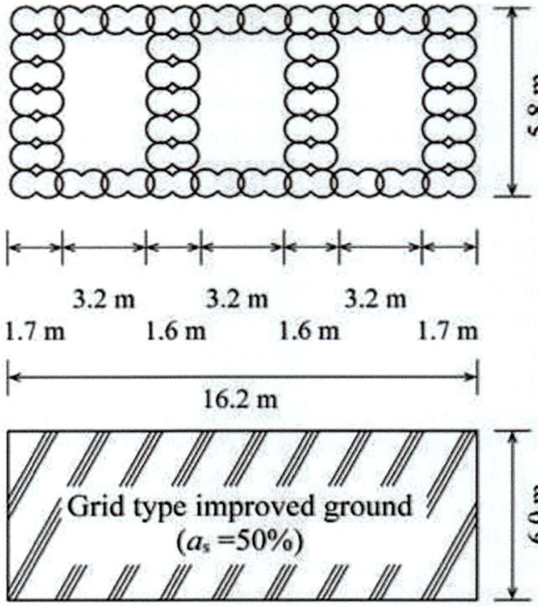

FIGURE 1.5 Plan and elevation views of DSM column layout in an embankment project in Soga, Chiba Prefecture.

(*Source:* Kitazume and Terashi 2013).

to the different requirements of the projects. In the case of the Bangna-Bangpakong Highway project as well as the approach embankment projects, the primary aim was to enhance bearing capacity and reduce embankment settlement. In such cases, the axial compressive load capacity of the columns is a key requirement. In the case of the embankment projects reported by Kitazume and Terashi (2013), containment of liquefied sand in the event of earthquakes is an important requirement. Hence, there is a necessity to ensure contiguity between columns.

Cement-treated soil columns have also been used to improve soil beneath large footing-type foundations, such as bridge abutments and wind turbine foundations (Topolnicki and Sołtys 2012; Topolnicki 2016). Steel I-sections were also placed in some of the cement-treated soil columns to permit them to sustain tensile forces arising from wind loading (e.g. see Figure 1.6). Andromalos et al. (2001) reported the use of short cement–soil columns to improve the improve the bearing capacity and reduce settlement of loose sandy soil underneath a storage tank. In all these instances, the cement-treated soil columns were installed as non-contiguous columns. This can be explained by the type of loading which cement-treated soil columns are subjected to. As Figure 1.7 illustrates, the loading on the soil columns ranges from axial compression in the interior of the loaded area to a combination of axial compression and shear near the edges of the loading area. In both of these loading scenarios, contiguity between treated soil columns is not a key requirement, whereas column axial capacity is.

FIGURE 1.6 Reinforced cement-treated soil columns for wind turbine foundation. **(*Source:* Topolnicki and Sołtys 2012).**

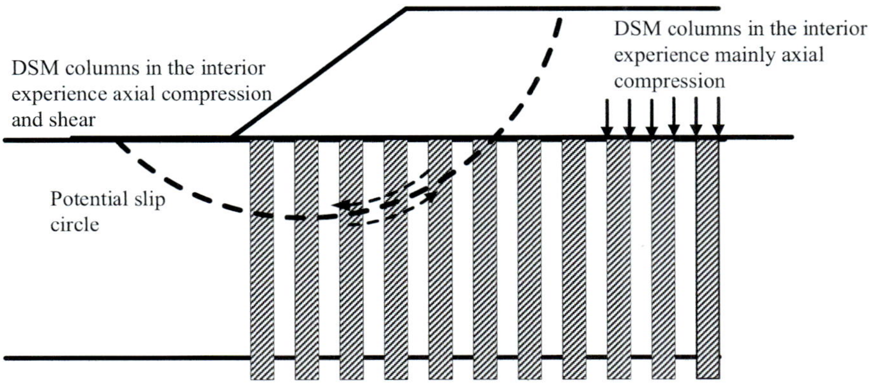

FIGURE 1.7 Loading on cement–soil columns (shaded) beneath an embankment

1.3.2 Seepage Cut-Off

Both ex and in situ chemical stabilisation approaches have been used to form cut-off walls to reduce seepage volume in levees, dams as well as excavations (e.g. Bruce and Sills 2009; Schwank 2016; Parashar et al. 2005; Cao et al. 2020). The ex situ approach typically consists of first excavating a trench under bentonite slurry, and then back-filling it using soil–cement or soil–cement–bentonite. Two in situ approaches are in common usage. The first consists of admixing the soil in situ with a mixture of cement and bentonite by DSM to form overlapping treated soil columns. Alternatively, instead of deep-mixed columns, contiguous cement–soil cut-off wall panels are formed using a chain saw slot-cutting equipment, such as a Trench Re-Mixing and Cutting Deep Wall (TRD) machine or a hydro-mill-type Cutter Soil Mixing (CSM) equipment (e.g. Bruce and Sills 2009). Regardless of the equipment used, the objective is to form a continuous cut-off wall of low permeability treated soil.

An alternative approach uses jet grouting to cut and mix soil in overlapping cylindrical columns (e.g. Croce and Modoni 2007; Burke 2007). This is commonly used as a remedial measure for existing dams as well as for seepage cut-offs in excavations, especially in congested urban environments (Cao et al. 2020). One of the reasons is that jet grouting uses high-pressure jets, instead of cutting blades, to cut the soil. Thus, the access opening in the ground required for jet grouting is much smaller than that for deep mixing; and the equipment is also generally smaller and lighter.

An important issue in seepage cut-off is the water-tightness of the cut-off wall, which imposes strict requirements on the relative alignment of the cement–soil or cement–soil–bentonite columns and wall panels (Greenwood 2009). Although there may be requirements on the strength and stiffness, these are usually less critical than water-tightness. This requirement is also reflected in common usage of bentonite in the admixture to lower permeability, especially in sandy soil (e.g. Bruce and Sills 2009) as well as to mitigate bleeding of water from the mix (e.g. Parashar et al. 2005), which may lead to void formation or loss of water-tightness. In some cases, soil–bentonite mix (without cement or binder) has also been used (e.g. Bruce and Sills 2009; Schwank 2016). Even where cement is used, the strength may not be a key concern (e.g. Parashar et al. 2005) unless there are other requirements for the cut-off wall, such as liquefaction containment, as in the case of the Jackson Lake Dam (Ryan and Jasperse 1989).

1.3.3 Excavation and Tunnelling Support

Chemical stabilisation has been widely used in underground construction since the 1980s. For instance, jet grouting (Tornaghi and Perelli Cippo 1985) and deep lime mixing (Hume et al. 1989) were used in the construction of the Singapore Mass Rapid Transit System Phase 1. Deep cement mixing has also been used for excavation support in situations where ground movement has to be strictly controlled, such as built-up urban environment (e.g. Tan et al. 2001; Li et al. 2011). Other instances of usage include the Boston Central Artery and Islais Creek Tunnel San Francisco (Pellegrino and Adams 1996), the Klang Valley Mass Rapid Transit in Malaysia (Yee and Tan 2015) as well as in Shanghai, China (Shao et al. 2005). Many other cases have also been reported.

One way of using chemical stabilisation is to incorporate the improved soil into part or whole of the retaining structure. Yee and Tan (2015) reported a case involving the use of DSM wall to support a 15-m deep shaft for tunnel boring machine removal at Taman Maluri, Kuala Lumpur in Malaysia (Figures 1.8 and 1.9). As Figure 1.8 shows, the Taman Maluri shaft has a rectangular configuration and is used to provide access to two tunnels. The three retaining walls, namely east, north and west walls, consisted of contiguous cement-stabilised soil columns forming three improved soil

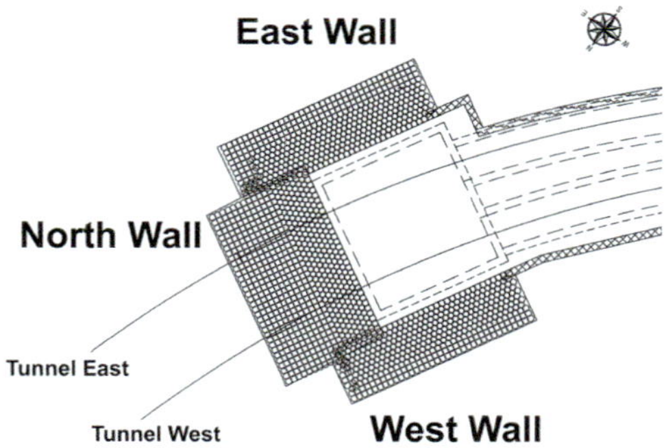

FIGURE 1.8 Cement-stabilised block used as retaining structure. Tunnel boring machine retrieval shaft at Taman Maluri, Kuala Lumpur, Malaysia.

(*Source:* Yee and Tan 2015).

FIGURE 1.9 DSM wall after tunnel boring machine exit, at Taman Maluri, Kuala Lumpur, Malaysia.

(*Source:* Yee and Tan 2015).

blocks, with wall thicknesses ranging from about 9m to about 17m. The relatively thick wall was used to prevent tensile failure of the improved soil. Another commonly used cement-stabilised column arrangement is that of grids (e.g. Topolnicki 2004, Leong et al. 2012). In addition, contiguous soil columns in an arch alignment have also been used, together with reinforced concrete bored piles, to form the retaining structure (Shao et al. 2005).

Cement-stabilised soil columns are also used to form improved soil layers below formation level, within the excavation (e.g. Page et al. 2006; Shirlaw et al. 2000a, 2000b, 2006) (Figure 1.10). The cement-stabilised soil slab enhances base stability

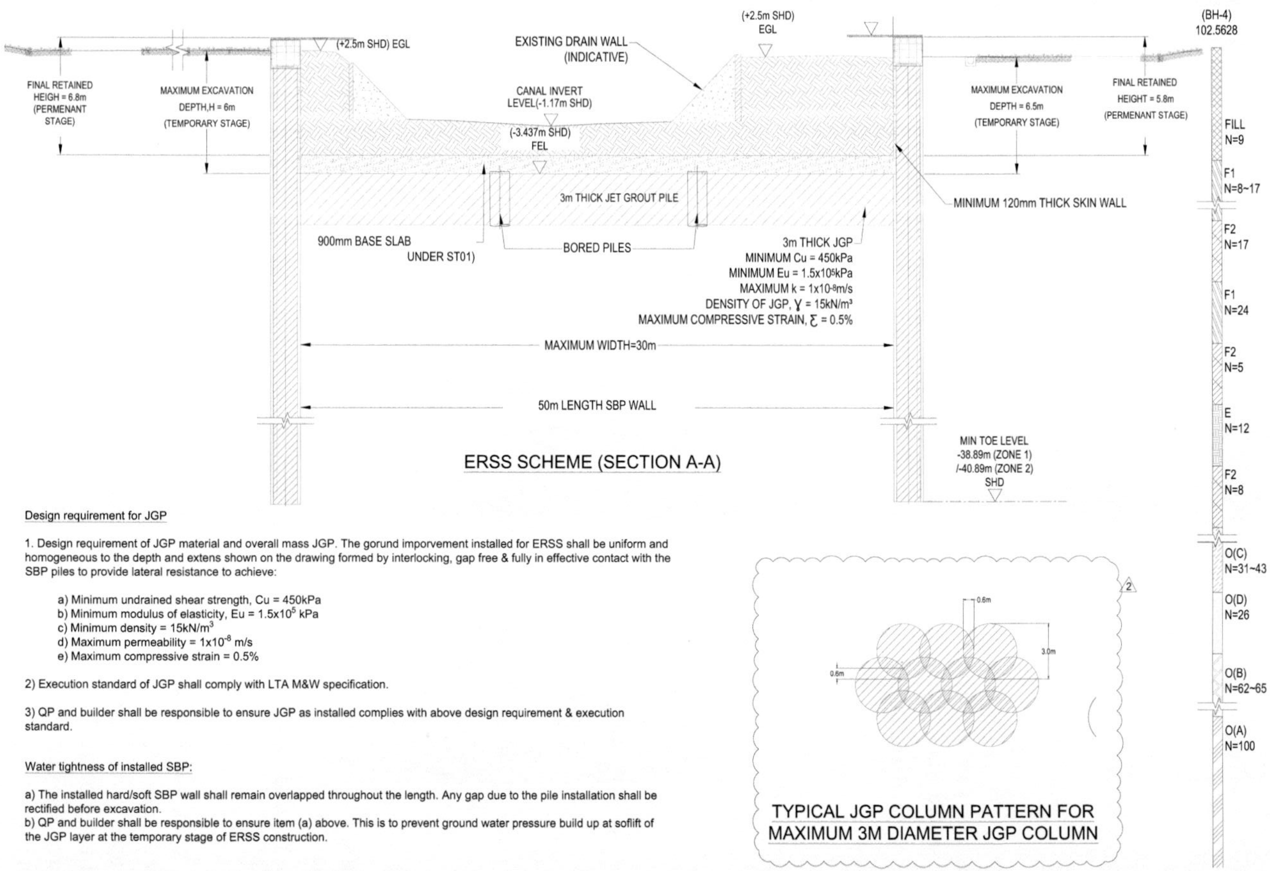

FIGURE 1.10 Schematic of a 3m thick jet-grout slab (region within black dashed line) installed below formation level to enhance base stability and limit wall deflection.

(*Source:* **Courtesy of ONE SMART Engineering).**

of the excavation and serves as an "underground strut" to limit inward deflection of the retaining wall. Compared to improved soil grids or blocks, it also minimises encroachment onto the surrounding areas behind the retaining walls; this can be an important consideration in highly built-up areas. Apart from the above, other arrangements have been employed. For instance, Tan et al. (2001) reported the use of contiguous lime-soil columns to form an improved soil berm to complement inclined struts in front of the retaining walls in a very wide excavation.

Chemical stabilisation has also been used to improve soft ground prior to tunnelling. In the Islais Creek Tunnel in San Francisco, the soft ground was jet-grouted to form a tubular improved soil surround enveloping the extrados of the tunnel to be dug (Pellegrino and Adams 1996). This permits tunnelling to be conducted using open-face shield, without compressed air. Jet grouting has also been conducted underground from within the tunnel. In many cases, this involves horizontal jet grouting to form a series of jet-grout umbrellas around the tunnel roof (e.g. Croce et al. 2004).

In such underground construction applications, strength and contiguity of soil columns are often important requirements. The strength is important since it affects the integrity and stiffness of the retaining structure or underground treated soil slab. This includes the tensile strength, which is closely related to the unconfined compressive strength (e.g. Terashi et al. 1980; Topolnicki 2004; Pan et al. 2016). Contiguity is also an important issue since the integrity of the improved soil mass relies closely on the transfer of stresses between adjacent columns. For the case of underground cement-treated soil slab, Liu et al. (2015) showed that a positioning error of a column that exceeds about 30% of the column radius would significantly degrade column-to-column contiguity, adversely affecting the mass strength and stiffness of the underground treated soil slab. This is particularly so in deep underground application, where even a small off-verticality may be sufficient to cause significant positioning error.

1.3.4 Roadworks

Chemical stabilisation has been also been applied to the treatment of road subgrade and subsoil. Often, the objectives of the treatment are to improve the strength of the underlying soft soil (Tatsuoka 2010) or to reduce heave-settlement amplitude in expansive soil conditions (e.g. Puppala et al. 2008). The improvement methodology usually takes the form of deep soil mixing, sometimes in conjunction with shallow soil mixing in the form of surface stabilisation. Puppala et al. (2008) noted that lime yields better results for treating expansive soil, compared to cement. Hausmann (1990) noted that chemical binders such as lime, cement and fly ash improve the soil in several ways. Firstly, the presence of quicklime reduces water content of the soil, leading to short-term strength gain. Secondly, lime also leads to reduction of soil plasticity leading to reduced heave potential. Finally, pozzolanic reaction leads to long-term strength gain. Cement has similar effects since cement hydration produces lime, albeit in smaller quantity for same binder weight. On the other hand, cement treatment usually confers more rapid hardening due to the short-term hydration reaction. Hence, cement is more often used for treating soft soil and loose sand whereas lime is more widely used for expansive soils.

Treatment typically takes the form of lime and cement columns, the chemical reaction in the surrounding soil being driven by the diffusion of the lime. The columns are often not contiguous. The axial strength and stiffness of the columns are important attributes if the objective of the treatment is to improve bearing capacity and reduce settlement of soft underlying soil. The requirements for chemical stabilisation for roadworks overlying soft ground are generally similar to those for foundation works. For reduction of heave potential, lime columns spaced at about 1m is typically employed (e.g. Puppala et al. 2008).

1.4 MIXING METHODOLOGY

The equipment for chemical stabilisation of soil varies substantially in detail, between manufacturers. However, the basic operating concepts of the equipment can be related to the improvement strategy.

1.4.1 Ex Situ Mixing

Ex situ mixing is often used where excavated or waste soil has to be re-deposited in a different location or requires additional treatment apart from stabilisation, such as decontamination. Ex situ mixing facilities typically consist of augers or excavators to remove soil from the site, batch or continuous mixing equipment as well as equipment for transporting the excavated soil from the site to the mixing equipment and from the mixing equipment back to the site or to another dump site. An example of an ex situ mixing method is the "liquefied soil stabilisation" method, which involves mixing cement and water with excavated soil to convert the latter into a self-compacting fluidised backfill and infill material for construction purposes (Figure 1.11) (e.g. Kuno et al. 1996; Miki et al. 2005; Onishi et al. 2005; Duong 2015).

The design of the soil mixing plant/equipment may vary depending upon the specific project requirements. However, they fall into two categories, namely batch and continuous mixing equipment. Batch soil mixing equipment is widely used in

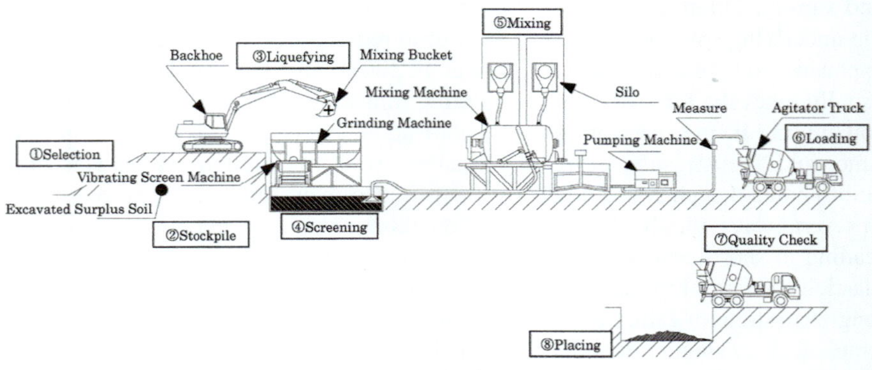

FIGURE 1.11 Typical work sequence of ex situ soil mixing.
(*Source:* Onishi et al. 2005; Duong 2015).

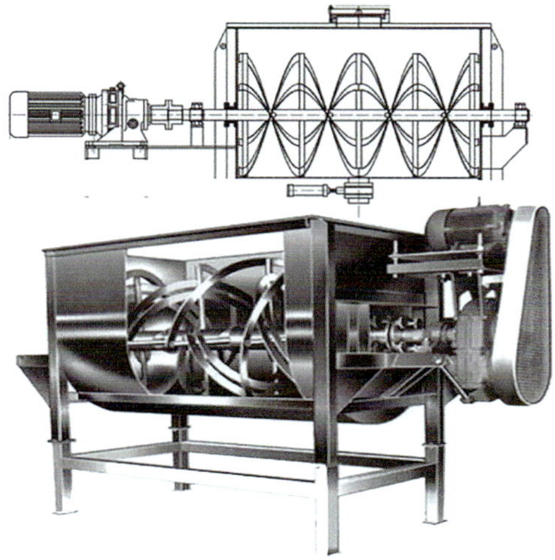

FIGURE 1.12 Horizontal ribbon blender/mixer for granular materials and powder. **(*Source:* Courtesy of JSK Engitech Industries).**

horticulture. Horticultural soil mixing equipment typically consists of a hopper placed on top of a ribbon blender (Figure 1.12). This type of mixer configuration is commonly used for granular material and powder mixing. However, ribbon mixers may not be suitable for soil stabilisation works, especially for cohesive, high-plasticity soil. where significant quantities of water often need to be added to transform the admixture into a viscous slurry to facilitate mixing. For such purposes, paddle mixers are usually the preferred configurations.

Most continuous mixers are based on ribbons and screw blenders, which transport and mix the components simultaneously. Continuous mixers have the advantage of a higher level of automation and probably higher throughput. However, their mixing efficacy is highly dependent upon the composition of the components. In soil stabilisation, there is often some variability in the components to be mixed. For this reason, batch mixers are often preferred in ex situ soil mixing. Apart from mixing, ex situ treatment may also involve decontamination processes such as leaching, washing and pyrolysis, if decontamination is also required.

1.4.2 In Situ Mixing

In situ mixing is usually the preferred approach for stabilising soil, which is to be reused on the same site, as it avoids the double handling required in excavating untreated soil and then re-depositing the treated soil. This is a particularly important factor if stabilisation is to be conducted at depths beyond about 5m. The basic aim of in situ mixing is to form hardened soil inclusions in the existing ground. These inclusions are most commonly in the form of cylindrical columns but can also be in the form of rectangular panels. Two approaches are often used to form these hardened soil inclusions, namely deep mixing and jet grouting.

1.4.2.1 Deep Mixing

The most commonly used method for deep soil mixing involves a rotary rig with auger and cutter blades mounted at the base of a hollow rotary shaft and mixing blades mounted further up the shaft (Figure 1.13). Nozzles are fitted at the rotary shaft, usually near the base, for the introduction of cementitious grout. The operations involved in deep mixing are illustrated in Figure 1.14. The shaft is rotated at a speed of typically between 12RPM and 70RPM while it is penetrated into the ground at the prescribed velocity to the desired depth. There are two main modes of introducing grout into the ground. The first mode, known as penetration injection method, introduces the grout during both penetration and withdrawal phase of the mixing shaft (Yoshizawa et al. 1997). The second, known as the withdrawal injection method, introduces the grout only during the withdrawal phase. In this mode, cementitious grout is pumped into the ground while the shaft is gradually withdrawn while rotated. The sequence of penetration and withdrawal may be repeated, if necessary, to improve the uniformity of the admixed soil. This produces a soil column consisting of an admixture of soil and cementitious grout. The RPM, penetration and withdrawal rates as well as grout flow rate are adjusted so that the prescribed grout content is achieved in the admixture and the desired degree of uniformity is reached. An operating parameter which is often used to control the mixing effort is the blade rotation number (CDIT 2002). This is defined as the number of mixing blade-times which each metre length of the stabilised soil column is subjected to during penetration and/or withdrawal mixing operation. More details on this parameter will be discussed in a later chapter.

Several variants of deep-mixing equipment are also in use. One of these involves mounting several rotary shafts onto a single rig, so that several overlapping columnar inclusions can be installed concurrently (Figure 1.15). This increases productivity by allowing multiple soil-mix columns to be formed in each penetration-withdrawal sequence.

Another version is the cutter soil mixing equipment shown in Figure 1.16. Instead of a vertical rotary shaft with augers and cutter bits attached, a cutter soil mixing head has multiple cutting disks which counter-rotate in a vertical plane. These cut and mix the soil in a deep trench instead of a cylindrical cavity. Cutter soil mixers are probably more productive than rotary shaft mixers for creating soil-mix walls and other long configurations. However, they are less widely used. As in the columnar deep mixing, there are also two main modes of grout injection (Bauer 2021). The first, known as the one-phase method, introduces the binder during the penetration phase. The second, known as the two-phase method, introduces bentonite and water during the penetration phase, followed by the binder during the withdrawal phase.

1.4.2.2 Jet Grouting

Another approach for admixing chemicals into soil is jet grouting. As Figure 1.17 illustrates, this approach uses high pressure water or grout jets to cut and mix the soil. A pilot cavity that is similar in diameter to that of a borehole is first drilled down to the soil layer to be improved. A grout tube in the form of a rotary shaft is then lowered down through the cavity. High pressure jets of grout or water are then used to cut the soil. Grout is also introduced into the cut cavity to harden the soil.

FIGURE 1.13 (a) Deep soil mixing rig and mixing shaft; (b) auger and mixing tool.
(*Source:* **Courtesy of Presscrete Engineering Pte. Ltd.**).

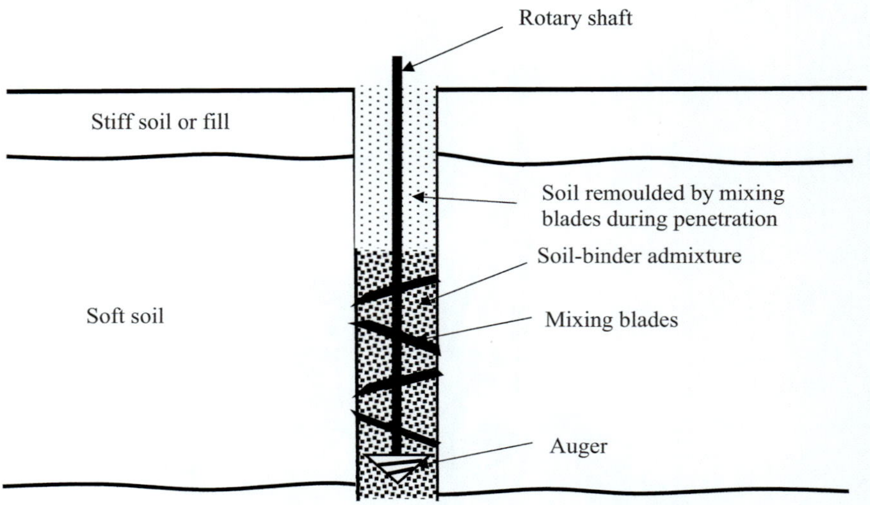

FIGURE 1.14 Schematic of deep soil mixing operation.

FIGURE 1.15 Deep soil mixing rig with triple mixing shafts.
(*Source:* Courtesy of Presscrete Engineering Pte. Ltd.).

FIGURE 1.16 Cutter soil mixing head.
(*Source:* Courtesy of Keller North America).

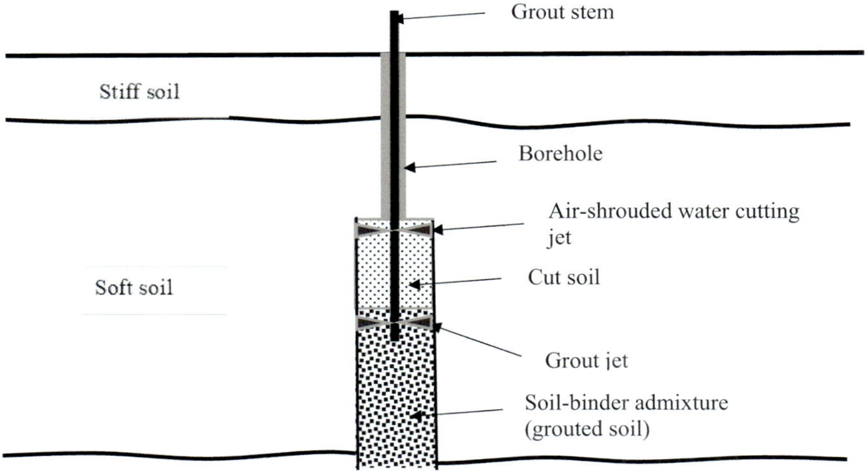

FIGURE 1.17 Triple tube jet-grouting sequence of operation.

Several variants of jet grouting exist. The simplest of these is single-tube or single fluid jet grouting, which uses a high-pressure grout jet to cut the ground. The grout is delivered in high-pressure form, typically 20MPa to 40MPa, through a hollow rotating shaft, and ejected through small nozzles with diameter ranging from about 1.5mm to 3mm. As the shaft rotates, the high-pressure grout jet cuts a cylindrical cavity in the soft soil. Another variant is double-tube or double-fluid jet grouting, which incorporates an additional air jet. The air and grout are delivered in concentric hollow shafts and air jet is ejected through an annulus around the grout jet. This allows the air jet to enshroud the grout jet, thereby mitigating resistance from the surrounding soil and slurry and increasing the cutting range of the grout jet. A third variant is triple-tube or triple-fluid jet grout, where air-shrouded water jets, instead of grout jet, are used to cut the soil. As water is less viscous than grout, higher nozzle velocity and longer cutting distance are achievable and lower grout pressure can also be used.

The diameter of the cut cavity typically ranges from about 1.5m to more than 5m, depending upon the cutting jet pressure as well as soil type and strength. Larger diameters are more likely to be achievable in sand than clay for the same standard penetration (SPT) blow count (N), as the high grout or water pressure can infuse into the sand matrix and erode it. Sandy soil with SPT N of 100 or even slightly higher may still be cut by the grout jet. On the other hand, clayey soils with SPT N greater than about 15 are often not treatable by jet grouting. In such cases, deep mixing is often preferred.

1.4.2.3 Deep Mixing or Jet Grouting?

Compared to deep mixing, an advantage of jet grouting is that the cutting and grouting jets are only turned on when the target depth is reached in the ground. Thus, a borehole-sized cavity is often sufficient to allow the grout stem to be lowered into the ground. In contrast, in deep mixing, the passage of the cutting blades will invariably create a much larger cylinder of disturbed soil above the layer to be treated. For this reason, jet grouting tends to produce less disturbance to the overlying soil. In situations where disturbance to overlying soil needs to be minimised, such as retrospective installation of cut-off walls in existing earth dams, jet grouting may be preferred.

The torque required to turn the jet-grout stem is also typically much lower than that for deep mixing since there is no necessity to overcome the soil resistance on the cutting blades. Jet-grouting equipment is generally less bulky and requires less headroom than deep-mixing equipment and may be preferred in situations where space and headroom are constrained.

Jet grouting is not a completely enclosed-volume process; the effluent from jet grouting, which is a mixture of water, air (for double- and triple-tube grouting), soil slurry and grout, is allowed to escape upwards through the annular gap between the borehole and the rotary shaft. Lee (1999) conducted a mass balance study of the composition of jet-grout effluent and proposed that jet grouting consists of mixing with partial replacement. A substantial proportion of the cut soil is expelled as effluent, together with portion of the grout and injected water, if any. In general, the effluent is leaner than the treated soil in terms of grout and richer in terms of cut soil. Lee et al. (2005) showed that jet-grouted soils generally have higher cement content and lower water–cement ratio than deep-mixed soils (Figure 1.18). This implies that higher strength may be potentially achievable with jet grouting than with deep mixing. However, this effluent needs to be collected and disposed off properly. In this sense, jet grouting involves greater amount of waste processing than deep mixing.

Both deep mixing and jet grouting have been observed to cause ground movement (Dong and Whittle 2018). In the case of deep mixing, the introduction of the

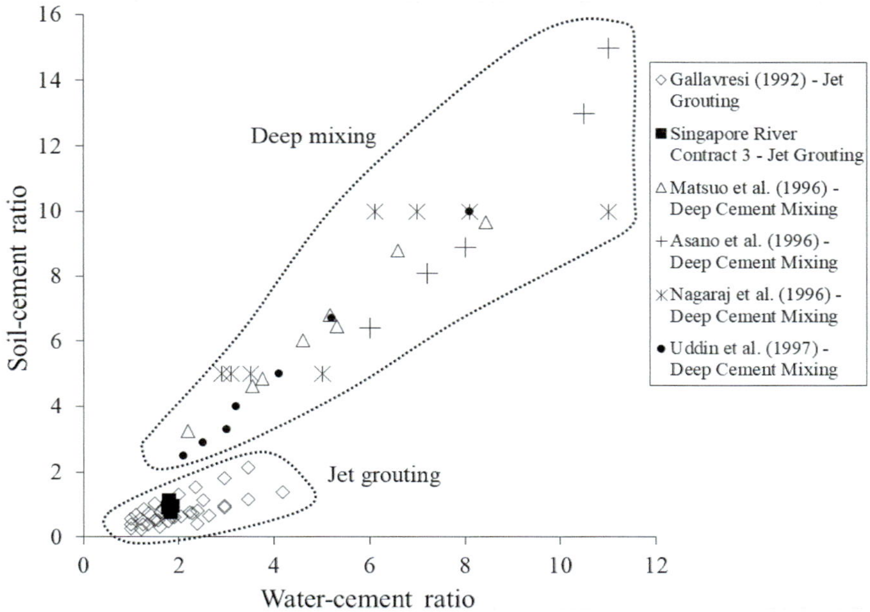

FIGURE 1.18 Soil–cement and water–cement ratios for some previous studies on deep mixing and jet grouting.

(*Source:* **Lee et al. 2005**).

grout into the ground will lead to an increase in the volume of the ground. Shen et al. (2006) noted that the diameter of the deep-mixed column is larger than that of the cut cavity due to the additional volume taken up by the grout. Lateral movement and heaving of the ground in and around the deep-mixed region have also been observed (e.g. Pan et al. 2022). This movement will accumulate and increase as successive columns are installed. If excessive, it may result in movement and damage to existing structures and foundations in the vicinity of the deep-mixing zone. Such mixing-induced ground movement tends to be more significant in situations wherein a large volume of soil is to be treated. To mitigate these effects, it may be advisable to commence the columnar mixing process at a location near to critical structures and then proceed away from the structure. This pushes the ground movement and heave wave away from the structures, rather than towards it, as the movement accumulates.

In many cases, jet grouting has been observed to cause larger ground heave than deep mixing in soft clay, due to the potential to build up higher pressure in the cut cavity. Jet grouting uses much higher pressure in its cutting jet than that used to deliver the grout in deep mixing. In some cases, the ground movement is attributed to blockage of the borehole by debris, leading to a build-up of high pressure in the jet-grout cavity, resulting in ground movement and heave during and after jet grouting (e.g. Shirlaw et al. 2003; Pinto et al. 2013; Cheng et al. 2023). However, Shirlaw et al. (2003) reported that significant heave may still occur in soft clay, even when the annular gap between borehole and grout shaft is not choked. They attributed this to the pressure required to balance the weight of the effluent up to the ground surface and the over-pressure required to move the viscous effluent upwards through the narrow annulus, which may be sufficient to cause lateral and upward expansion of the cut cavity. Finally, jet grouting may not be as effective as deep mixing in cutting through stiffer soils.

In some equipment, such as RAS Jet system (Raito 2007), the principles of deep mixing and jet grout are combined by incorporating jet-grouting nozzles at the ends of the cutting/mixing tools. This increases the radius of the cut cavity beyond that of the cutting/mixing tools and allows larger columns to be installed.

1.4.2.4 Shallow Soil Mixing

Shallow soil mixing equipment is usually designed to mix soil in shallow locations rather than large depths. Early shallow soil mixing equipment were modified from the deep-mixing concept; they were mounted on drilling or rotary rigs, but with shorter rotary shaft and large diameter auger. However, modern shallow soil mixing equipment more commonly take the form of modified cutter soil mixer mounted on an excavator arm. An example of this is the mass soil mixing equipment shown in Figure 1.19. The cutting and mixing bits are mounted on a cylindrical cutterhead, which is in turn mounted on an arm which can be attached to an excavator. During the mixing operation, the cutterhead is aligned horizontally, so that the rotation is in a vertical plane. Instead of a rotary shaft, the rotational motion is rotated by a hydraulic motor fitted within the cutterhead, with the hydraulic pipes running through the mounting arm. The excavator arm allows the cutterhead to be moved vertically as well as horizontally.

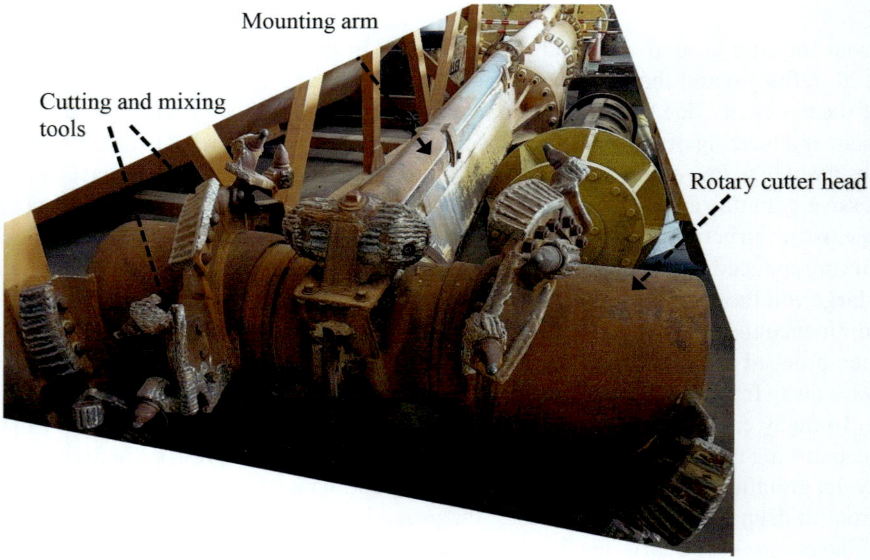

FIGURE 1.19 Mass soil mixing equipment.
(*Courtesy:* **Keller Foundations SEA Pte Ltd**).

Shallow mixing does not produce well-defined columnar or wall-like structures such as those produced by rotary shaft and cutter soil deep mixing. Instead, the soil is mixed en masse. Although some form of control may still be applied by subdividing the area to be treated into panels, control parameters, which can be used to assess mixing effort, and thus mixing quality, such as the blade rotation number used in rotary shaft mixing, are still not well established. For this reason, mixing uniformity may be more uncertain than shallow soil mixing.

REFERENCES

Andromalos, K. B., Hegazy, Y. A. and Jasperse, B. H. (2001). Stabilization of soft soils by soil mixing. *Geotechnical Special Publication 112—Soft Ground Technology*: 194–205. https://doi.org/10.1061/40552(301)16.

Bauer (2021). *CSM cutter soil mixing*. BAUER Maschinen GmbH, CSM 905.656.2 7.

Bergado, D. T., Ruenkrairergsa, T., Taesiri, Y. and Balasubramaniam, A. S. (1999). Deep soil mixing used to reduce embankment settlement. *Ground Improvement*, 3: 145–162.

Bruce, D. A. and Sills, G. (2009). Seepage cut-offs for levees: A technology review. *29th Annual USSD (United States Society of Dams) Conf.*, Nashville, Tennessee: 969–998.

BS EN 14679 (2005). *Execution of special geotechnical works—deep mixing*. British Standards Institution.

Burke, G. K. (2007). Vertical and horizontal groundwater barriers using jet grout panels and columns. *Proc.* In *Geo-Denver 2007* 1–10. https://doi.org/10.1061/40912(231)8.

Cao, C., Shi, C., Liu, L. and Liu, J. (2020). Evaluation of the effectiveness of an alternative to control groundwater inflow during a deep excavation into confined aquifers. *Environmental Earth Sciences*, 79: 502, 13p.

CDIT (2002). *The deep mixing method—principle, design and construction*. Coastal Development Institute of Technology, A. A. Balkema Publishers, Lisse/Abingdon/Exton (PA)/Tokyo, Japan.

Cheng, S. H., Chao, K. C., Wong, R. K. N. and Wang, M. I. M. (2023). Control of jet grouting process induced ground displacement in clayey soil. *Transportation Geotechnics*, 40: 100983.

Croce, P. and Modoni, G. (2007). Design of jet grouting cut-offs. *Proceedings of the Institution of Civil Engineers—Ground Improvement*, 11(1): 11–19.

Croce, P., Modoni, G. and Russo, G. (2004). Jet-grouting performance in tunnelling. *GeoSupport 2004*, ASCE, Orlando, FL: 910–922.

Dong, Y. P. and Whittle, A. J. (2018). Effects of deep soil mixing and jet grouting on adjacent retaining walls and ground. In *Geotechnical aspects of underground construction in soft ground* (Eds. A. Negro and M. O. Cecilio Jr.), Taylor & Francis Group, London: 91–96. ISBN: 978-1-138-29711-1.

Duong, Q. H. (2015). *Study on application to backfilling ground by liquefied stabilized soil mixed with fibered material in Hanoi city*. PhD thesis, Muroran Institute of Technology Academic Resources Archive.

Fleming, R., Sills, G. and Stewart, E. (1992). Lime stabilization of levee slopes. *2nd Interagency Symposium on Stabilization of Soils and Other Materials*, Los Angeles: 5-15–5-22.

Greenwood, D. A. (2009). Discussion: Design of jet-grouting cut-offs. *Proceedings of the Institution of Civil Engineers—Ground Improvement*, 162(G11): 47–50.

Gutschick, K. A. (1978). Lime stabilization under hydraulic conditions. *4th Lime Congress*, Hershey, Philadelphia: 1–20.

Hausmann, M. R. (1990). *Engineering principles of ground modification*, McGraw-Hill, Singapore.

Howard, A. K. and Bara, J. P. (1976). Lime stabilization on Friant-Kern canal. *U.S. Bureau of Reclamation, Report No. REC-ERC-76-20*, Denver, CO: 61p.

Hume, T. W., Potter, L. A. C. and Shirlaw, J. N. (1989). Singapore mass rapid transit system construction. *Proceedings of the Institution of Civil Engineers*, 86(4): 709–770.

Kitazume, M. and Terashi, M. (2013). *The deep mixing method*, CRC Press/Balkema, EH Leiden, The Netherlands.

Kuno, G., Miki, H., Mori, N. and Iwabuchi, J. (1996). Application of the liquefied stabilized soil method as a soil recycling system. *Proc. 2nd Int. Congress Environmental Geotechnics* (Ed. M. Kamon), Osaka, Japan: 791–796.

Lee, F. H., Lee, Y., Chew, S. H. and Yong, K. Y. (2005). Strength and modulus of marine clay-cement mixes. *Journal of Geotechnical and Geoenvironmental Engineering*, 131(2): 178–186.

Lee, Y. (1999). *A framework for the design of jet grout piles in Singapore marine clay.* MEng dissertation, National University.

Leong, W. K., Soh, K. K., Yeo, L., Chew, S. H., Leong, K. W. and He, Z. W. (2012). Deep soil mixing columns as retaining structure. *18th South East Asian Geotechnical Conference*, Singapore: 157–164.

Li, G. J., Wong, K. S. and Ng, P. B. (2011). Back analysis of a braced excavation with DCM ground improvement. *Proc. Underground Singapore 2011,* Singapore: 85–93.

Liu, Y., Lee, F. H., Quek, S. T., Chen, E. J. and Yi, J. T. (2015). Effect of spatial variation of strength and modulus on the lateral compression response of cement-admixed clay slab. *Geotechnique*, 65(10): 851–865.

Liver, N. L. (1962). Mixed in place pile. *United States Patent No. 3023585*: 6p.

Miki, G. (1973). Chemical stabilization of sandy soils by grouting in Japan. *8th Int. Conf. Soil Mech. & Found. Engrg.*, Moscow, Russia: 395–405.

Miki, H., Iwabuchi, J. and Chida, S. (2005). New soil treatment methods in Japan. *Treatment and Recycling of Materials for Transport Infrastructure (TREMTI) 2005*, Communication C189.

Miller, M. M. (2000). Lime. In *U.S. geological survey minerals yearbook—2000*, US Geological Survey: 46.1–46.7.

Nakanishi, W. (1974). Method for forming an underground wall comprising a plurality of columns in the earth and soil formation. *United States Patent 3,800,544*: 8p.

Onishi, T., Nozu, M., Yoshitomi, H., Fujii, M. and Akishige, H. (2005). Liquefied stabilized soil method for building foundation. *Japanese Society of Material Science*, 54(11): 1129–1134 (in Japanese).

Page, R. J., Ong, J. C. W., Osborne, N. and Shirlaw, J. N. (2006). Jet grouting for excavations in soft clay—design and construction issues. *Proc. Int. Conf. Deep Exc.*, Singapore.

Pan, H., Tong, L., Wang, Z. and Yang, T. (2022). Effects of deep soil mixing on existing shield tunnels in soft soil ground. *Underground Space*, 7: 724–733.

Pan, Y. T., Xiao, H. W., Lee, F. H. and Phoon, K. K. (2016). Modified isotropic compression relationship for cement-admixed marine clay at low confining stress. *Geotechical Testing Journal ASTM*, 39(4): 20150147.

Parashar, S. P., Wong, I. H. and Sanmuganathan, D. (2005). Ground water cut-off wall at Changi water reclamation plant project. *Underground Singapore*: 79–88.

Pellegrino, G. and Adams, D. N. (1996). The use of jet grouting to improve soft clays for open face tunnelling. In *Geotechnical aspects of underground construction in soft ground* (Eds. R. J. Mair and R. N. Taylor), Balkema, Rotterdam. ISBN: 90-5410 856-8.

Pinto, F., O'Neil, G. M., Apfelbaum, A. D. and Whittle, A. J. (2013). Ground heave due to jet grouting near an existing structure. *Proc. 7th Annual Conference on Case Histories in Geotech. Engrg, Chicago, IL*. Missouri Uni. Sc. & Tech., Paper No. 2.53.

Puppala, A. J., Madhyannapu, R. S., Nazarian, S., Yuan, D. and Hoyos, L. (2008). Deep soil mixing technology for mitigation of pavement roughness. *FHWA/TX-08/0-5179-1*, Federal Highway Administration, USA: 342p.

Radić, J., Kindij, A. and Mandić, A. (2008). History of concrete application in development of concrete and hybrid arch bridges. *Chinese-Croatian Colloquium on Long Arch Bridges*, Brijuni Islands, Croatia, 10–14 July: 9–118.

Raito (2007). *RAS-JET system—mechanical soil mixing combined with jet grouting*. https://www.raito.co.jp/english/construction/pdf/ras-jet.pdf.

Raito (2018). *Shallow Soil Mixing (SCM)—rotary blender and bucket mixing*. https://www.raito.co.jp/english/construction/pdf/scm.pdf.

Ryan, C. R. and Jasperse, B. H. (1989). Deep soil mixing at the Jackson Lake Dam. *ASCE Specialty Conference, Found. Engrg.: Current Principles and Practices* (Ed. F. H Kulhawy), ASCE, New York, NY: 354–367.

Schwank, S. (2016). Cut-off wall technologies in mining. *IMWA (International Mine Water Association) 2016* (Eds. C. Drebenstedt and M. Paul), Freiberg, Germany: 468–477.

Shao, Y., Macari, E. J. and Cai, W. (2005). Compound deep soil mixing columns for retaining structures in excavations. *Journal of Geotechnical and Geoenvironmental Engineering*, 131(11): 1370–1377.

Shen, S. L., Pang, X. M. and Gu, W. H. (2006). Analysis of the diameter of deep mixed column. *Proc. GeoShanghai 2006: Ground Modification and Seismic Mitigation, Shanghai* (Eds. A. Porbaha, S. L. Shen, J. Wartman and J. C. Chai), GSP 152, Geo-Institute, American Society of Civil Engineers, Reston, VA, USA: 17–22.

Shirlaw, J. N., Tan, T. S. and Wong, K. S. (2006). Deep excavations in Singapore marine clay. *5th Int. Conf. Geot. Aspects Underground Construction Soft Grd.* (Eds. K. J. Bakker, A. Bezuijen, W. Broere and E. A. Kwast), CRC Press, Amsterdam, Netherlands: 13–28.

Shirlaw, J. N., Wen, D., Kheng, H. Y. and Osborne, N. (2003). Controlling heave during jet grouting in Marine Clay. *Rapid Transit Conference Singapore 2003*, Singapore: 1–10.

Shirlaw, J. N., Wen, D., Nadarajah, P., Yoon, S. I. and Sugawara, S. (2000a). Construction issues related to jet grouted slabs at the base of excavations. *Int. Conf. Tunnels and Underground Structures*, Singapore: 639–643.

Shirlaw, J. N., Wen, D., Nadarajah, P., Yoon, S. I. and Sugawara, S. (2000b). Design issues related to jet grouted slabs at the base of excavations. *Int. Conf. Tunnels and Underground Structures*, Singapore: 653–658.

Tan, T. S., Ng, T. G., French, D., Wong, F. H. and Takeda, T. (2001). Use of an improved soil berm for stabilisation in deep excavation. *Proc. Underground Singapore 2001*, National University of Singapore, Singapore: 367–377.

Tatsuoka, F. (2010). Cement soil mixing for highway improvement. *Soils and Foundations*, 50(6): 785–804.

Terashi, M., Tanaka, H., Mitsumoto, T., Niidome, Y. and Honma, S. (1980). Fundamental properties of lime and cement treated soils. *2nd Report, 19(1), Port and Harbour Research Institute*, Yokosuka, Japan: 33–62 (in Japanese).

Topolnicki, M. (2004). In situ soil mixing (Chapter 9). In *Ground improvement* (Eds. M. P. Moseley and K. Kirsch), Spon Press, New York.

Topolnicki, M. (2016). General overview and advances in deep soil mixing. *XXIV Geotechnical Conference of Torino Design, Construction and Controls of Soil Improvement Systems*, Torino: 1–30.

Topolnicki, M. and Sołtys, G. (2012). Novel application of wet deep soil mixing for foundation of modern wind turbines. *4th International Conference on Grouting and Deep Mixing*, New Orleans, USA: 10p.

Tornaghi, R. and Perelli Cippo, A. (1985). Soil improvement by jet grouting for the solution of tunnelling problems. *Proc. Tunnelling '85* (Ed. M. J. Jones), Institute of Mining and Metallurgy: 265–283.

Yee, Y. W. and Tan, Y. C. (2015). Excavation support for TBM retrieval shaft using deep soil mixing technique, Kuala Lumpur. *Int. Conf. and Exhibition on Tunnelling and Underground Space*, Petaling Jaya, Malaysia.

Yoshizawa, H., Okumura, R., Hosoya, Y., Sumi, M. and Yamada, T. (1997). JGS TC report: Factor affecting the quality of treated soil during execution of DMM. *Proc. 2nd Int. Conf. on Ground Improvement Geosystems*, Tokyo, 2: 931–937.

2 Chemical Basis of Soil Stabilisation

2.1 BACKGROUND

Stabilisation of soil using chemical binders has been studied and has been used extensively in various civil engineering works since the 1970s. These include road and highway construction, stabilisation of marginal land with soft and contaminated soils, land fill and reclamation as well as excavations and underground construction. The chemicals that are most widely used include cement, lime, fly ash, ground granulated blast furnace slag (GGBS). Other binders and binder components include sulphoaluminate cement (e.g. Chen et al. 2020), gypsum (e.g. Zhu et al. 2018; Chen et al. 2018, 2020; He et al. 2021), cement kiln dust (e.g. Miller and Zaman 2000), bentonite (e.g. Katsioti et al. 2008; Lin et al. 2009) and mining waste products such as jarosite and alunite (e.g. Boura et al. 2003; Katsioti et al. 2009). Some of these are used as components in a cocktail of binders. Additives are sometimes required to predispose the conditions to initiate as well as to balance the stoichiometry for the hardening reactions. Non-cementitious binders such as asphaltic emulsion (Lucena et al. 2014) have also been studied. However, as they have not been used on a wide scale, they will not be discussed herein.

Chemical treatment of clay soils often involves two types of hardening reactions. The first is a hydration reaction, which involves only the binder and water. The second is a pozzolanic reaction, which involves the lime within or produced by the binders, water with silica and alumina.

2.2 HYDRATION REACTION

A hydration reaction is a chemical reaction in which a substance reacts with water. Common soil stabilisation chemicals, which undergo hydration reaction, are quicklime or calcium oxide (CaO), CEM I Ordinary Portland Cement (OPC) or blended cement, such as fly ash cement (CEM II), or blast furnace slag cement (CEM III), which contains OPC as a component.

Quicklime undergoes hydration reaction with water to produce hydrated lime $Ca(OH)_2$ through the following reaction:

$$CaO + H_2O \rightarrow Ca(OH)_2 \tag{2.1}$$

Since hydrated lime does not produce cementation, its production does not confer any strength gain. However, the resulting lowering of water content in the soil matrix may result in an increase in strength and this has been utilised in some applications where the water content of the soil is excessive (e.g. Xue and Chen 2015). In

DOI: 10.1201/9781003285885-2

addition, hydrated lime can react with pozzolans to produce cementitious products. This will be discussed further. Apart from quicklime, calcium carbide, which undergoes hydration reaction with water to produce acetylene and hydrated lime, has been used for chemical stabilisation work (e.g. Rafalko et al. 2007).

OPC is a mixture of four chemical compounds, known as bogue compounds. These are alite or tricalcium silicate (short form: C_3S), belite or dicalcium silicate (C_2S), tricalcium aluminate (C_3A) and tetracalcium aluminoferrite (C_4AF). Hence, its hydration reaction is also a mix of different reactions, most of which have variable stoichiometry. The most abundant cementitious component in OPC is tricalcium silicate (C_3S), which makes up roughly 60% of OPC and contributes much of the strength development in cement. The hydration reaction of C_3S can be written as follows:

$$C_3S + (3 - y + x)\,H \rightarrow C_y\text{-}S\text{-}H_x + (3 - y)CH \quad (2.2)$$

Here, C, S and H stand for calcium oxide, silicate and water, respectively, C_y-S-H_x being the cement chemist notation for calcium silicate hydrate (C-S-H) phase, the hyphenation and variables x and y denoting variable stoichiometry. The reaction product CH is calcium hydroxide, or hydrated lime. The variable x represents the amount of water associated with the C-S-H gel, which varies from about 1.4 to 4, depending on the water content in the admixture and how much of the water is considered to be part of C-S-H. The variable y is the ratio between calcium oxide and silicon dioxide in C-S-H and typically varies between 1.2 and 2.1. Much of the reaction occurs during the first few days, leading to substantial early strength gains and reduction in capillary porosity.

The second most abundant cement compound is dicalcium silicate (C_2S), which makes up about 13% to 20% of OPC. This compound reacts with water according to the following reaction:

$$C_2S + (2 - y + x)\,H \rightarrow C_y\text{-}S\text{-}H_x + (2 - y)CH \quad (2.3)$$

The hydration products are the same as those of C_3S, but the amount of hydrated lime formed is less than that of C_3S. C_2S is much less soluble than C_3S, so the rate of hydration is much slower. C_2S hydration contributes little to the early strength of cement but makes substantial contributions to the strength of mature cement paste and concrete.

Tricalcium aluminate (C_3A) and tetracalcium aluminoferrite (C_4AF) occur together in even smaller proportions than C_2S alone. Their combined proportion is typically not more than 20%. The hydration reaction of tricalcium aluminate (C_3A or $Ca_3Al_2O_6$) produces calcium aluminate hydrate, but not lime. Calcium aluminate hydrate occurs in several forms and is produced by reacting 1 molecule of Al_2O_3 with 1 to 6 molecules of quicklime, the most common form being $m(CaO)\cdot Al_2O_3\cdot nH_2O$, with m and n denoting variable stoichiometry

All hydration reactions consume water. According to Neville (1995), the water–cement mass ratio needed for complete hydration of a cement mix is approximately 0.42:1. However, during the hydration process, gel water is also produced. The net mass of water consumed by the complete hydration reaction is approximately 0.23 times the mass of cement solids. Neville (1995) noted that the volume of the cementitious products of hydration, including gel water, is approximately 74.6% of the total volumes of the dry cement and water consumed during hydration. In other words, there is volumetric loss of approximately 25% of the volume of dry cement and water consumed.

Cement hydration requires an alkaline environment. In concrete and many soils, this is not a problem because cement contains highly soluble alkali oxides (Na_2O and K_2O), which, when dissolved, will produce the high pH, typically 12 to 13, in the pore solution required to initiate the hydration reaction. Once the hydration reaction is underway, the production of hydrated lime, $Ca(OH)_2$, will help to sustain the alkaline environment. However, Tremblay et al. (2002) noted that some compounds can strongly hinder the hardening process by inhibiting hydration reactions. In particular, they noted that "organic acids, mixed with soil and cement that produce a pH lower than 9 in the pore solution, prevent the development of the cementing products because the pH is too low to allow secondary mineral formation" (p. 545).

Some soils such as peat, as well as undigested or partially digested sewage sludge used as landfill, contain substantial proportion of organic matter. The presence of such organic content may, in some cases, affect the efficacy of treatment using cement or cementitious compounds. For example, Kang et al. (2017) noted that a threshold level of cement is required to overcome the adverse effect of humic acid on strength development in cement-treated clay. Soils with significant organic content often have significant levels of humic acid. For instance, the humic acid content in peat typically ranges from about 1% to about 10% by mass (Wilkomirski and Malawska 2004). Sludge also has similar levels of humic acid (e.g. Li et al. 2014; Kulikowska et al. 2019). Chen and Wang (2006) and Morse et al. (2007) noted the tendency of black humic acid to react with calcium ions to form calcium humate. This binds up calcium ions from lime and renders them unavailable for pozzolanic reaction.

Calcium sulphoaluminate cement is cement containing calcium sulphoaluminate ($Ca_4(AlO_2)_6SO_4$), or ye'elimite, as a main constituent. In cement chemist notation, it is represented by C_4A_3s, in which $A = Al_2O_3$, $C = CaO$ and s = SO3. It is often regarded as a promising low-carbon (CO_2) alternative to OPC. Gypsum is also usually included to enable the formation of the cementitious products ettringite, the chemical name for which is calcium aluminium sulphate, $Ca_6Al_2(SO_4)_3(OH)_{12}{\cdot}26H_2O$ or C_3A 3Cs 32H, and strätlingite, $Ca_2Al_2SiO_7{\cdot}8H_2O$ or C_2ASH_8. The chemical name for gypsum is calcium sulphate dihydrate $CaSO_4.2H_2O$ or CsH_2. The hydration reactions may be represented as follows:

$$C_4A_3s + 2CsH_2 + 34H \rightarrow C_3A\ 3Cs\ 32H + 2AH_3 \tag{2.4}$$

$$\text{and } C_2S + AH_3 + 5H \rightarrow C_2ASH_8 \tag{2.5}$$

Reaction (2.4) produces ettringite and aluminium hydroxide while reaction (2.5) uses calcium silicate and aluminium hydroxide to produce strätlingite, also known as calcium aluminium silicate hydrate. In contrast to OPC, the hydration reaction of calcium sulphoaluminate cement produces aluminium hydroxide instead of hydrated lime ($Ca(OH)_2$).

2.3 POZZOLANIC REACTION

The pozzolanic reaction is often the reaction sustaining long-term gain in strength in cement–soil treatment. This is a chemical reaction that occurs between hydrated lime and a broad class of chemicals known as pozzolans, in the presence of water. It is the main reaction utilised in Roman concrete, which was used to build the Pantheon. The main pozzolans are silica (SiO_2) and alumina (Al_2O_3). Silica is the main chemical component of quartz sand, but sand is typically unable to sustain pozzolanic reactions since it has a low ratio of particle surface-to-volume ratio, owing to its large particle size. For the same reason, sand does not participate in the hardening reaction in concrete. Hence, in cement treatment of sand, the predominant hardening reaction is the hydration reaction. The pozzolanic reaction converts a silica- or alumina-rich precursor with no cementing properties, to calcium silicate hydrate and calcium aluminate hydrate, which are cementitious products.

The pozzolanic reaction occurs between hydrated lime ($Ca(OH)_2$), and silicic acid (written as $(SiO_2)\cdot(H_2O)_x$, where x lies between 0.5 and 3. A typical reaction is as follows:

$$3Ca(OH)_2 + 2[(SiO_2).\ (H_2O)_x] \rightarrow [3(CaO).\ 2(SiO_2).(2x+3)(H_2O)] \quad (2.6)$$

where $[3(CaO).\ 2(SiO_2).(2x+3)(H_2O)]$ is usually denoted as C-S-H. In the notation of cement chemists,

$$3CH + 2SH_x \rightarrow C\text{-}S\text{-}H \quad (2.7)$$

The ratio of calcium oxide to silicon dioxide and the number of water molecules can vary. Thus, the cementitious products of the pozzolanic reaction are similar to those obtained from hydration reaction.

Many pozzolans may also contain aluminate, or $Al(OH)_4^-$, which will react with calcium hydroxide and water to form calcium aluminate hydrates (C-A-H) such as C_4AH_{13}, C_3AH_6 or hydrogarnet. In combination with silica, strätlingite or calcium aluminium silicate hydrate can also be produced. In addition, smaller quantities alumina-ferric-oxide-monosulphate, also known as aluminate ferrite monosulphate (AFm), and ettringite may also be produced. However, the main cementitious products are C-S-H and strätlingite.

The conditions for pozzolanic reaction are similar to those of hydration reaction, namely free water and an alkaline environment. Davidson et al. (1965) noted that pH must be higher than 10.5 to initiate and maintain the pozzolanic reaction. This is because a high pH is required to ensure that silicon and aluminium ions are sufficiently soluble to sustain the pozzolanic reaction. Leonard (1958) also noted that the pH tends to decrease during pozzolanic reaction as lime is being consumed. Hence, the initial pH should preferably be 12 or higher to ensure that pH does not fall below ~10.5 as lime is consumed. The establishment of an alkaline condition, termed alkali-activation, is also used in geopolymerisation involving calcined clay.

In general, the pozzolanic reaction is slower than the hydration reaction. Whereas the latter will start to produce the hardening effect within a matter of hours, the former will usually require more than 7 days. For the case of OPC, it often takes an even longer time since the hydration reaction has to reach a certain point before the concentration of lime is sufficient to sustain the pozzolanic reaction. If additional lime is introduced, pozzolanic reaction will be able to initiate earlier. Nonetheless, the pozzolanic reaction remains slower than the hydration reaction. Hence, binders based on pozzolanic reactions will typically require longer than 28 days to reach full strength.

2.4 REACTIONS OF OPC WITH SOFT CLAY MINERALS

Many binders have been proposed for use in soil improvement. However, one of the most commonly used binders is OPC, which produces hardening through hydration and pozzolanic reactions. The pozzolanic reaction induced by OPC comes from the lime produced by it. Hence, by examining the hardening process of OPC in the presence of pozzolanic soil minerals, it is possible to understand many of the hardening processes of cementitious chemicals.

Chew et al. (2004) examined the physico-chemical reaction between OPC and a high plasticity soft marine clay, which occurs widely in Singapore, which contains kaolin and illite as the main clay minerals. Their findings indicate that calcium ions from the hydrated lime induce a flocculated clay structure, which comprises clay clusters separated by large intra-cluster voids, filled with water. The increase in effective size of the particles and the presence of entrapped water cause an almost equal increase in the plastic and liquid limits. In addition, the permeability of the soil may also be increased. Kaolinite is preferentially consumed, compared to illite, in the pozzolanic reaction. This is consistent with high pozzolanic potential of kaolinite arising from its hydroxyl groups (Fernandez et al. 2011). Illite was not completely consumed by the pozzolanic reaction after 28 days of curing. Chew et al. (2004) suggested that the pozzolanic reaction could have stopped due to encapsulation of the illite by the cementitious products, thereby rendering the illite inaccessible to the lime. However, it is also possible that illite and montmorillonite may require longer duration to undergo pozzolanic reaction. Nonetheless, this suggests that the pozzolanic reaction may not necessarily continue until pozzolans are exhausted. Other factors, such as exhaustion of lime and physical encapsulation of the pozzolanic material by cementitious products, may also lead to cessation of the pozzolanic reactions.

In contrast to OPC, binders such as lime and GGBS rely more exclusively on pozzolanic reaction. Lime requires an adequate supply of pozzolans to produce cementitious products. Thus, lime is only effective on its own, as a binder, if there is an adequate amount of pozzolans in the soils. Otherwise, lime should be used in conjunction with other pozzolanic ingredient. GGBS, however, contains lime, silica and alumina and therefore can sustain its own pozzolanic reaction. Hence, consideration must be given to the availability of ingredients when selecting binders.

2.5 INGREDIENTS AND MIX PROPORTIONS FOR POZZOLANIC REACTION

2.5.1 Variability in Binder Types and Mix Proportions

In chemical stabilisation of soil, the mix proportion is commonly quantified in terms of water (C_w) and binder (A_w) contents, which are defined by the following:

$$C_w = \frac{w}{s+c} \tag{2.8a}$$

$$A_w = \frac{c}{s} \tag{2.8b}$$

in which c and s are the dry masses of binder and soil (or sludge), respectively, and w is the total mass of water at the point of mixing.

The mix proportion can also be deduced from the in situ state of the soil and the amount of the water and binder added to it. Let w_0 be in situ water content of the soil before treatment and W_1 be the mass of water added during stabilisation. Then

$$w = w_0 s + W_1 \tag{2.9}$$

For stabilisation involving wet mixing, the water–binder (or water–cement) ratio a of the slurry is usually specified. In such cases, $W_1 = ac$ and Equations (2.9) and (2.8a) can be expressed, respectively, as follows:

$$w = w_0 s + ac \tag{2.10}$$

$$C_w = \frac{w_0 + aA_w}{1 + A_w} \tag{2.11}$$

Table 2.1 summarises the ingredients and the relative amounts of binder components in some soil improvement studies involving pozzolanic reaction. As can be seen, the binder types, mix proportions and strength vary widely between the different studies. Moreover, the efficacy of different binders with different soils (or sludge) appears to differ substantially. For instance, Chew et al. (2004), using cement content of 60% and water content of 90%, obtained a 28-day strength of about 1050kPa for soft inorganic marine clay. On the other hand, He et al. (2021), using binder content of 413% and water content of 85.8%, obtained a much lower 28-day strength of about 220kPa, even though the water content is similar.

This illustrates an important challenge facing chemical stabilisation works; that is improved strength depends upon numerous influencing factors, such as the following:

1. Soil mineralogy or sludge composition
2. Binder type
3. Binder content
4. Water content

To date, there is still no analytical framework which allows the binder and mix proportion to be estimated, given the soil and its in situ state. Instead, trial mixes are

TABLE 2.1

Ingredients used in some soil stabilisation studies.

Authors	Soil Type	Binder(s)	Results
Chew et al. (2004)	Marine clay. PI = 52%, LL = 87%. In situ water content ~70%. Mineralogy: illite 18%, quartz = 71%, kaolinite 11%.	OPC. Cement content = 10%–60%. Water content = 90% and 120%.	28-day strength ranges from 150kPa to 1050kPa
Lee et al. (2005)	Marine clay. PI = 59%, LL = 90%.	OPC. Cement content = 25%–213%. Water content = 54%–160%	28-day strength ranges from 100kPa to 850kPa.
Uddin et al. (1997)	Marine clay. PI = 60%, LL = 103%. In situ water content ~76%–84%.	OPC. Cement content = 5%–40%.	
Lin et al. (2009)	Dewatered wastewater sludge. PI = 256%, water content = 314%–357% (this is the water content of the raw sludge). Organic content = 42.8%.	Sludge:OPC:bentonite = 1:0:0.5 to 1:0.4:0.4. Binder content = 50%–80%	For OPC content = 40% and bentonite content = 40%, i.e. binder content 80%, strength ranges from ~12kPa–~500kPa.
Zhu et al. (2018)	Dewatered sewage sludge. Water content = 78%.	OPC and modified steel slag, which consists of steel slag, activator (consisting mainly of Al_2O_3, SiO_2 and CaO) and gypsum. Binder content ~20%.	With 20% modified steel slag, 14-day strength = 74.5kPa With OPC, 14-day strength = 29.5kPa.
Chen et al. (2020)	Sewage sludge. Water content = 83%, organic content = 38%.	Proprietarybindersulphoaluminate-based cement:OPC:CaO: gypsum:lithium salt = 0.30:0.60:0.05:0.049:0.001. Binder content = 118%.	(a) Initial water content = 56%, treated strength ~580kPa. (b) Initial water content = 85%, treated strength ~60kPa.
He et al. (2021)	Sewage sludge. LL = 380%, PL = 63%, water content = 566%. SiO_2 45%, Al_2O_3 16.4%, CaO 5.6%. Sludge has a shortage of CaO.	Sludge solids:soda residue:GGBS: CaO = 1:2–3.33:1.33:0.8. Binder content = 413%–546%. Water content = 85.8%–68.2%.	(a) For mix ratio 1:2:1.33:0.8, binder content 413% and water content 85.8%, strength ~220kPa. (b) Mix ratio 1:2.66:1.33:0.8, binder content 479% and water content 82.1%, strength ~280kPa. (b) Mix ratio 1:3.33:1.33:0.8, binder content and water content 68.2%, strength ~410kPa.

$$\text{Cement (or binder content)} = \frac{\text{Dry mass of cement (or binder)}}{\text{Dry mass of soil (or sludge)}} \times 100\%$$

Strength refers to unconfined compressive strength.

PI and LL stand for plasticity index and liquid limit, respectively.

usually conducted at the start of the project to determine the effective binder types and mix proportions. In cases where previous experience with similar soil types is available, workable binder types and mix proportions may be known approximately. Nonetheless, some amount of trial and error is still generally required.

2.5.2 Binder–Soil Stoichiometry

In spite of the variable stoichiometry, examining the stoichiometric balance of the hardening processes may still allow us to understand binder–soil reaction and provide some guidelines for estimating mix proportions in binder–soil reaction. In this section, we will examine binder–soil reaction quantitatively and attempt to derive some estimates of mass proportions. A basic assumption made herein is that the main pozzolanic reactions take place between lime and silica/alumina, or minerals containing them, such as kaolinite, which contains silica and alumina sheets in their molecular structure. This assumption is not strictly true since not all silica and alumina are pozzolanic. For instance, sand cannot sustain pozzolanic reactions because of its large particle size. Illite and montmorillonite, on the other hand, are less pozzolanic than kaolinite even though they also contain silica sheets. This is because they contain fewer hydroxyl groups. Thus, their crystalline structures are less readily decomposed (Fernandez et al. 2011). Finally, encapsulation by cementitious products may render some of the pozzolans unavailable for the pozzolanic reaction.

The mass ratios of the reactants and products involved in the hydration and pozzolanic reactions can be roughly estimated from the chemical reactions. Table 2.2 shows the atomic and molecular weights of some common substances.

2.5.2.1 Hydration Reaction

2.5.2.1.1 Tricalcium Silicate (C_3S)

If we assume that x = y = 1.5 in Reaction (2.2), then the hydration reaction for tricalcium silicate is given by the following:

$$2(3CaO{\cdot}SiO_2) + 6H_2O \rightarrow 3CaO{\cdot}2SiO_2{\cdot}3H_2O + 3Ca(OH)_2 \qquad (2.12)$$

Expressing the stoichiometric balance in terms of molecular weight,

$$2 \times (3 \times 56.077 + 60.083) + 6 \times 18.015 \rightarrow \{3 \times (56.077) + 2 \times (60.083) + 3 \times 18.015\} + 3 \times (74.092) = 564.718$$

In other words,

456.628 mass-units of C_3S + 108.09 mass-units of water → 342.442 mass-units of calcium silicate hydrate + 222.276 mass-units of hydrated lime.

Normalising the mass-units by those of C_3S, 1 mass-unit of C_3S will produce about 0.49 mass-units of hydrated lime and about 0.75 mass-unit of C-S-H.

TABLE 2.2
Approximate atomic/molecular weights of some common compounds (after IUPAC 2021).

Name	Atom/Molecule	Approximate Atomic/ Molecular Weight
Calcium	Ca	40.078
Silicon	Si	28.085
Aluminium	Al	26.981
Hydrogen	H	1.008
Oxygen	O	15.999
Water	H_2O	18.015
Quicklime (calcium oxide)	CaO	56.077
Silica (silicon dioxide)	SiO_2	60.083
Alumina (aluminium oxide)	Al_2O_3	101.959
Hydrated lime (calcium hydroxide)	$Ca(OH)_2$	74.092
	CEM I Ordinary Portland Cement	
Tricalcium silicate (alite)	Ca_3SiO_5	228.314
Dicalcium silicate	Ca_2SiO_4	172.237
	Cementitious Products	
Calcium silicate hydrate (product of hydration reaction)	$3CaO{\cdot}2SiO_2{\cdot}3H_2O$ [$2(C_{1.5}$-S-$H_{1.5})$]	342.442
	$CaO{\cdot}SiO_2{\cdot}H_2O$ [C-S-H]	134.175
Calcium aluminate hydrate*	$3CaO{\cdot}Al_2O_3{\cdot}6H_2O$ [Tricalcium aluminate hexahydrate or C_3-A-H_6]	378.28
	$2CaO{\cdot}Al_2O_3{\cdot}8H_2O$ [Dicalcium aluminate octahydrate or C_2-A-H_8]	358.233
	$CaO{\cdot}Al_2O_3{\cdot}10H_2O$ [Calcium aluminate decahydrate or C-A-H_{10}]	338.186

* There are several variants of calcium aluminate hydrate but the main strength-giving variants at room temperature are calcium aluminate decahydrate (C-A-H_{10}) and dicalcium aluminate octahydrate (C_2-A-H_8) (Bensted 2008).

The specific mass of the products will depend on the values of x and y. Another common form of calcium silicate hydrate is obtained by setting x = y = 1 in Reaction (2.2), leading to the following:

$$3CaO{\cdot}SiO_2 + 3H_2O \rightarrow CaO{\cdot}SiO_2{\cdot}H_2O + 2Ca(OH)_2 \tag{2.13}$$

For this reaction, 1 mass-unit of C_3S will produce about 0.65 mass-units of hydrated lime and about 0.59 mass-unit of C-S-H.

Finally, assuming $x = 2$ and $y = 1.5$ leads to the following:

$$2(3CaO{\cdot}SiO_2) + 7H_2O \rightarrow 3CaO{\cdot}2\,SiO_2{\cdot}4H_2O + 3Ca(OH)_2 \tag{2.14}$$

In this case, 1 mass-unit of C_3S will also produce about 0.49 mass-units of hydrated lime and about 0.79 mass-units of C-S-H. Based on the aforementioned results, we can estimate that 1 mass-unit of C_3S will produce about 0.5 to 0.65 mass-units of hydrated lime and about 0.6 to 0.8 mass-units of C-S-H upon hydration.

2.5.2.1.2 Dicalcium Silicate

The hydration reaction for dicalcium silicate (C_2S) can be similarly estimated. Assuming $x = y = 1.5$ leads to the following:

$$2(2CaO{\cdot}SiO_2) + 4H_2O \rightarrow 3CaO{\cdot}2SiO_2{\cdot}3H_2O + Ca(OH)_2 \tag{2.15}$$

For this reaction, 1 mass-unit of C_2S will produce about 0.22 mass-unit of hydrated lime and approximately 1 mass-unit of C-S-H. Hence, dicalcium silicate produces much less hydrated lime than tricalcium silicate since it has less calcium oxide.

Assuming that each mass-unit of OPC has about 0.6 and 0.2 mass-units of C_3S and C_2S, respectively, it can be shown that each mass-unit of OPC will produce roughly 0.3 to 0.4 mass-units of hydrated lime and roughly 0.55 to 0.67 mass-units of C-S-H. The lime will only be available if there is complete hydration. If hydration is incomplete, lime production will be lower than estimated.

2.5.2.2 Pozzolanic Reaction

The stoichiometry for the pozzolanic reaction between lime, silica and water can be similarly estimated. From Reaction (2.6), we obtain the following:

$$\begin{aligned} &3\times(56.077+18.015)+2\times60.083+2x\times18.015 \rightarrow 3\times(56.077)+2\times(60.083) \\ &+(3+2x)\times18.015 \rightarrow \underbrace{168.231}_{\textit{Lime}}+\underbrace{120.166}_{\textit{silica}}+\underbrace{(3+2x)\times18.015}_{\textit{water}} \end{aligned} \tag{2.16}$$

Normalising by the molecular weight of silica, about 1.4 mass-units of lime (calcium oxide) will completely react with 1 mass-unit of silica to produce C-S-H. The amount of water in the C-S-H and thus its mass-unit depends on value of x, but this does not alter the ratio of lime to silica. For $x = 0$ to 3, the masses of C-S-H lie between 2.85 and 3.75 mass-units, respectively, for every mass-unit of reacted silica.

To form calcium aluminate hydrate, $Ca_2AlO_3(OH){\cdot}nH_2O$, via pozzolanic reaction, we need to react lime with alumina and water. The typical reaction is given by the following:

$$mCaO + Al_2O_3 + nH_2O = mCaO{\cdot}Al_2O_3{\cdot}nH_2O \tag{2.17}$$

The total mass-units of C-A-H produced is $56.077m + 101.959 + 18.015n$.

The values of m and n typically lie between 1 to 3 and 6 to 10, respectively. At room temperature, n is typically 10, which produces calcium aluminate decahydrate. This implies that, for complete reaction between lime and alumina, we typically require a lime-to-alumina mass ratio of between ~0.55 and ~1.65. For n = 10 and m between 1 and 3, 1 mass-unit of alumina produces about 3.3 to 4.4 mass-units of C-A-H. The mass of C-A-H, which can potentially be produced per unit mass of lime, ranges from about 2.7 to 6.

Kaolinite comprises silica and alumina in high percentage. Illite and montmorillonite, on the other hand, also contain other cations such as potassium, magnesium and iron. Weaver and Pollard (1973) reported that well-organised illite contains approximately 9% to 10% potassium oxide. Some illite may also contain magnesium and iron as substitutes for aluminium in the octahedral sheet (Marshall 1964). Similarly, in montmorillonite, 1 out of every 6 Al^{3+} cations is substituted by a Mg^{2+} cation. However, these percentages are generally small and most illites and montmorillonite contain about 83% to 90% silica and alumina. The molar ratio of silica to alumina in kaolinite, illite and montmorillonite is about 2:1, 2.5:1 and 4:1, respectively. This gives a mass ratio of roughly 1.2:1, 1.5:1 and 2.4:1, respectively. The mass ratio for kaolinite agrees well with Salahudeen and Idris' (2022) value of 1.25, while that for illite lies within Lindgreen et al.'s (1991) reported range of 1.39 to 2.02. Table 2.3 shows the estimated mass ratios for lime–kaolinite reaction under ideal conditions, assuming that kaolinite comprises entirely of silica and alumina. If all the silica and alumina in kaolinite are available as pozzolans, the mass ratio of lime to kaolinite to achieve a balanced reaction, that is one in which both lime and pozzolans are exhausted, ranges from about 1.0 to 1.5.

Such ideal conditions may not always exist. For instance, if OPC is used as the binder, then cementitious products from the hydration reaction may encapsulate some of the soil particles, thereby rendering the latter inaccessible to the lime (e.g. Chew et al. 2004; Guo et al. 2022). This is likely to occur with high OPC content,

TABLE 2.3
Mass ratios for lime–kaolinite pozzolanic reaction.

Constituents	Mass-Unit: Minimum Lime	Mass-Unit: Maximum Lime
Kaolinite	1.000	1.000
Silica:alumina mass ratio	1.200	1.200
Silica	0.545	0.545
Alumina	0.455	0.455
Lime		
Lime to react with silica ~1.4 times the mass-unit of silica	0.764	0.764
Lime to react with alumina ~0.55 to 1.65 times the mass-unit of alumina	0.250	0.750
Total lime	1.014	1.514

TABLE 2.4
Typical silica, alumina and lime contents in IBA, GGBS and Ordinary Portland Cement.

	Percentage by Mass (%)			
Binder	Silica	Alumina	Lime Needed to React with Silica and Alumina	Actual Amount of Lime
Incineration bottom ash (IBA)	~38	~10–17	~58–81	20–23
Ground granulated blast furnace slag (GGBS)	~29	~13	~48–62	~44

which results in large amounts of cementitious compounds being produced by hydration reaction. On the other hand, if lime is used as the sole stabilising agent, then the absence of cementitious products from the hydration reaction may permit more silica and alumina to react with lime. Unreacted lime or pozzolan may not only contribute no strength but its presence in the soil as a soft, uncemented component may degrade the strength of the treated soil.

Some binders contain both lime and pozzolans. As Table 2.4 illustrates, two examples of such binders are incinerator bottom ash (IBA) and ground granulated blast furnace slag (GGBS). Based on the aforementioned calculations, it can be shown that IBA does not contain sufficient lime to react with all of its own silica and alumina. GGBS is also slightly deficient but is much closer to the stoichiometric balance. Hence, on its own, GGBS is a more balanced cementing agent. If the soil also has some silica and alumina, then addition of lime into these two binders may improve their efficacy. It should also be noted that, in its raw form, IBA often takes the form of very coarse aggregates, which cannot sustain pozzolanic reaction. It has to be grounded into a very fine powder form.

2.5.3 Activation Threshold

A minimum binder content is often required before there is some stabilisation effect. For instance, Chew et al.'s results (2004) showed at OPC content of 5% or less, the gain in strength of the treated marine clay is insignificant. Kang et al. (2017) also noted that a threshold level of cement is required to overcome the adverse effect of humic acid on strength development in cement-treated clay. Similarly, in GGBS treatment, a small amount of lime or alkaline material (~5%) is often required to hydrolyse the solid reactants and activate the reaction (e.g. Palomo et al. 1999; Dung et al. 2019). All these imply that a minimum binder content is often required to initiate the production of cementitious compounds. The existence of this threshold can be attributed to the fact that both the hydration and the pozzolanic reactions require a minimum binder content to raise the pH to the required level.

For soils with significant organic contents, the activation threshold is likely to be higher. Figure 2.1 shows the increase in undrained shear strength of treated sludge with lime and OPC content. The partially digested sludge is a waste product from waste

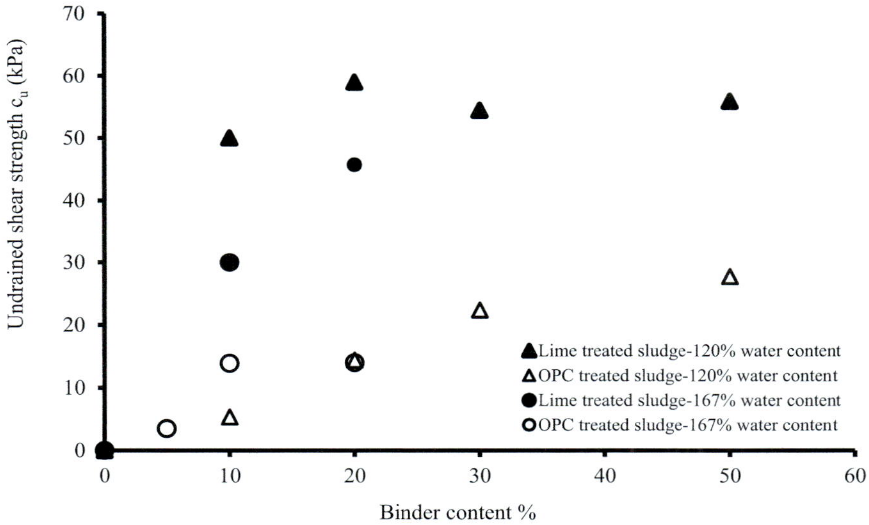

FIGURE 2.1 Effect of Lime and OPC Content on a Treated Partially Digested Sludge

water treatment that has been buried in sandy ground for about 10 to 20 years. It consisted of soft blackish clayey clumps with an average diameter of 100mm to 200mm. The in situ water content is around 120%. Its bulk unit weight is about 13kN/m^3. The particulate composition is approximately 68% sand and 32% fines. The plastic limit (PL) and liquid limit (LL) are approximately 100% and 150%, respectively, indicating a highly plastic material. Although organic content was not measured, it is expected to be significant given its origin. As Figure 2.1 shows, the overall gain in strength for lime and OPC content up to 50% is relatively modest. This is much lower than the undrained shear strength of about 300kPa to 400kPa for inorganic soft clay (Chew et al. 2004; Lee et al. 2005). Moreover, the strength does not pick up until the binder content exceeds about 10%; this is higher than the threshold value of 5% reflected in Chew et al.'s (2004) results on inorganic marine clay. This indicates that soils with significant organic content, such as peaty clay and sludge used as landfill, may be more resistant to chemical stabilisation than inorganic soils.

Tremblay et al. (2002) also noted that a pH higher than 11.5 may be required to initiate hydration reaction in soils with high organic content. They attributed this to the adverse effect of humic acid, which is commonly present in organic soil and digested sludge, and which has the effect of lowering the pH value. Moreover, Chen and Wang (2006) and Morse et al. (2007) noted that black humic acid tends to react with calcium ions to form calcium humate. This binds up calcium ions, rendering them unavailable for pozzolanic reaction. Pertusatti and Prado (2007) noted that humic acid also has pH buffering capability for pH ranging from 5.5 to 8, with the buffering effects derived from chemisorption of hydroxyl groups. This implies that a threshold level of alkaline compounds is required to offset the effect of the humic acid and raise the pH to the level required to initiate the hydration or pozzolanic reaction.

The existence of a threshold activation level of lime required to initiate the hydration or pozzolanic reaction may be represented by a "threshold lime content" ζ, the value of which can be back-deduced from test results using different lime contents. OPC has small amounts of magnesium oxide (about 1% to 3%) and other alkaline materials, which, in sufficient quantity, can also raise the pH to the required level.

The value of ζ can be estimated from the gain in strength as binder content is increased. In Figure 2.1, the threshold lime (and OPC) content is about 10%. However, 68% by weight of the particles of this sludge are sand, which is likely to have been mixed from the surrounding soil during the landfilling process. Only 32% are silt-sized or finer. If we assume that the active components of the sludge are contained within the fine-grained portion while the sand is essentially an inert component, then we may normalise the threshold lime content by the proportion of fines to give the threshold lime content of the fines. Doing so gives an approximate threshold lime content of about 31%. In other words, if the fine-grained portion is separated from the sand, then the threshold lime content will be roughly 30% before the strength gain can initiate.

Figure 2.2 shows the results from Boura et al. (2003) and Malliou et al. (2007) on the strength of fresh sludge from wastewater treatment works, after stabilisation. The

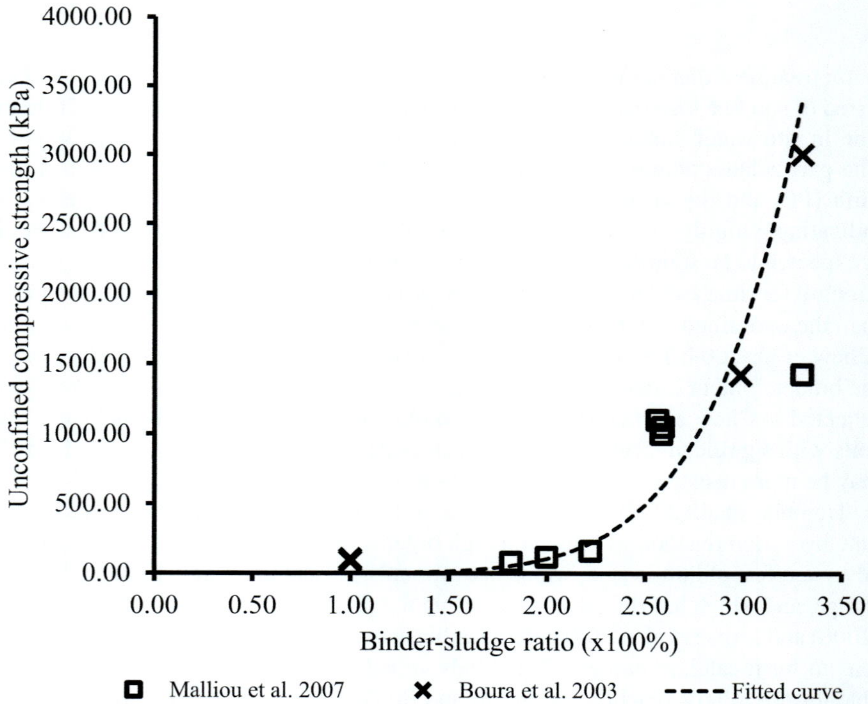

FIGURE 2.2 Strength of sludge against binder content. Binder comprises mainly OPC with small quantities of jarosite, alunite, lime and calcium hydroxide.

sludge was pre-mixed with sand with sand content ranging from 85% to 95%. The binder consists mainly of OPC with smaller amounts of jarosite, alunite, lime and calcium chloride. As Figure 2.2 shows, the threshold binder–sludge ratio (in percentage) is of the order of 250%. This is much higher than the threshold value implied by Figure 2.1 and indicates that there is significant variability in the properties of sludge. In particular, the sludge referred to in Figure 2.1 is a partially digested sludge whereas Boura et al.'s (2003) and Malliou et al.'s (2007) data pertain to fresh sludge. As such, some form of testing is advisable.

2.5.4 A Framework for Understanding Strength Gain by Chemical Stabilisation

The notions of stoichiometric balance and activation threshold permit the strength gain of soft clays and sludge due to chemical treatment to be understood quantitatively, albeit only approximately. An example is Chew et al.'s (2004) and Lee et al.'s (2005) studies on the strength gain of soft, high plasticity marine clay stabilised using OPC. The marine clay has liquid limit and plasticity index of about 88% and 55%, respectively, and is classified as an inorganic high plasticity clay. The cement content used in their studies ranges from 10% to 213%. The hydration reaction requires only water to proceed, but the pozzolanic reaction requires pozzolans and lime to be available. Hence, stoichiometric balance should be satisfied to ensure that pozzolanic reaction can proceed.

Table 2.5 shows the stoichiometric balance sheet for the lime required to react with the pozzolans in the soils and that generated by different binder contents, as well as the cementitious products. This balance can be represented by the following relationship:

$$\left(\lambda_2 P_2 + \lambda_3 P_3\right) A_w = T_a + \left(\lambda_s P_s + \lambda_a P_a\right) W_p \quad (2.18)$$

P_2 and P_3 are the proportions of C_2S and C_3S in OPC, taken herein to be 0.6 and 0.2, respectively,
λ_2 and λ_3 are the masses of lime generated by hydration of unit mass of C_3S and C_2S, respectively,
T_a is the activation threshold mass of lime needed to initiate the pozzolanic reaction,
λ_s and λ_a are the masses of lime required to fully react with unit mass of pozzolanic silica and alumina, respectively,
P_s and P_a are the proportions of silica and alumina, respectively, in kaolinite,
W_p is the mass of pozzolan, taken herein to be kaolinite, per unit mass of soil solids.

The right-hand side of Equation (2.18) represents the amount of lime required to fully react with the silica and alumina in the soil. As Table 2.5 shows, the mass of lime required in this example is 0.189 per unit mass of soil. The left-hand side

TABLE 2.5

Stoichiometric balance in lime–silica and lime–alumina reactions in OPC and soft marine clay.

Lime Demand in Pozzolanic Reaction			
Soil	**Marine clay**	**Marine clay**	
Soil Pozzolan	**Kaolinite**	**Kaolinite**	**Remarks**
Mass of pozzolan/mass of soil solids, W_p (assumed to be kaolinite)	0.11	0.11	Chew et al. (2004) noted that kaolinite is exhausted after 28-days of curing, but amount of illite remains almost unchanged. Kaolinite is thus assumed to be the only pozzolan.
		Silica	
Proportion of silica in kaolinite, P_s	0.545	0.545	Based on a ratio of 1.2:1 for silica:alumina mass ratio in kaolinite.
Mass of reactable silica/mass of soil solids = W_pP_s	0.0600	0.0600	
Mass of lime needed to react with unit mass of silica λ_s	1.4000	1.4000	Based on the estimate that about 1.4 mass-units of calcium oxide will completely react with 1 mass-unit of silica
Mass of lime needed to react with silica/mass of soil solids = $\lambda_s W_p P_s$	0.0840	0.0840	
Mass of C-S-H potentially producible by available silica			
Mass of C-S-H produced/mass of silica, ζ_s	3.30	3.30	This is the median value of the estimated 2.85 to 3.75 mass-units of C-S-H produced per mass-unit of silica.
Mass of C-S-H produced/mass of soil solids, $\zeta_s P_s W_p$	0.20	0.20	
		Alumina	
Proportion of alumina in kaolinite, P_a	0.455	0.455	Based on a ratio of 1.2:1 for silica:alumina mass ratio in kaolinite.
Mass of reactable alumina/mass of soil solids = W_pP_a	0.0500	0.0500	
Mass of lime needed to react with unit mass of alumina, λ_a	1.1000	1.1000	This is the median value of the estimated 0.55 to 1.65 mass-units of lime needed to react with the alumina.
Mass of lime needed to react with alumina/mass of soil solids = $\lambda_a W_p P_a$	0.0550	0.0550	
Mass of C-A-H potentially producible by available alumina			
Mass of C-A-H produced/mass of alumina, ζ_a	3.85	3.85	This is the median value of the estimated 3.3 to 4.4 mass-units of C-A-H produced per mass-unit of alumina.

Lime Demand in Pozzolanic Reaction			
Soil	**Marine clay**	**Marine clay**	
Soil Pozzolan	**Kaolinite**	**Kaolinite**	**Remarks**
Mass of C-A-H produced/mass of soil solids, $\zeta_a P_a W_p$	0.19	0.19	
Total mass of C-S-H + C-A-H produced by pozzolanic reaction/mass of soil solids (**pozzolan-limited**)	**0.39**	**0.39**	
	Activation threshold		
Mass of lime needed to initiate pozzolanic reaction/mass of soil solids, T_a	0.0500	0.0500	Assumed activation threshold = 5%.
Total mass of lime needed to fully react with pozzolans/ mass of soil solids, $= T_a + \lambda_s W_p P_s + \lambda W_p P_a$	**0.1890**	**0.1890**	
	Lime supply from hydration reaction		
Binder	OPC	OPC	
Binder content $100A_w$ (%)	25.00	50.00	
Mass of binder/mass of soil solids, A_w	0.25	0.5	
Proportion of C_3S in OPC, P_3	0.6	0.6	
Proportion of C_2S in OPC, P_2	0.2	0.2	
Mass of lime generated by hydration of unit mass of C_3S, λ_3	0.575	0.575	This is the median value of the estimated 0.5 to 0.65 mass-unit of lime generated by hydration of C_3S.
Mass of lime generated by hydration of unit mass of C_2S, λ_2	0.22	0.22	
Total mass of lime generated by hydration reaction/mass of soil solids = $(\lambda_2 P_2 + \lambda_3 P_3)A_w$	**0.097**	**0.195**	
Total mass of C-S-H produced by hydration reaction/mass of binder, ψ_h	0.61	0.61	This is the median value of the estimated 0.55 to 0.67 mass-unit of C-S-H produced by OPC hydration.
Total mass of C-S-H produced by hydration/mass of soil solids, $A_w \psi_h$	0.1525	0.305	
	Amount of C-S-H or C-A-H potentially producible by available lime		
Fraction of lime assumed to react with silica, χ	0.545	0.545	The proportions assumed here follow the proportion of silica and alumina in kaolinite.
Fraction of lime assumed to react with alumina $(1-\chi)$	0.455	0.455	

(Continued)

TABLE 2.5 *(Continued)*
Stoichiometric balance in lime–silica and lime–alumina reactions in OPC and soft marine clay.

Lime Demand in Pozzolanic Reaction			
Soil	**Marine clay**	**Marine clay**	
Soil Pozzolan	**Kaolinite**	**Kaolinite**	**Remarks**
Mass of C-S-H produced/mass of lime, $\zeta_{ls} = \zeta_s/\lambda_s$	2.357	2.357	
Mass of C-A-H produced/mass of lime, $\zeta_{la} = \zeta_a/\lambda_a$	3.500	3.500	
Total mass of C-S-H + C-A-H produced by pozzolanic reaction/mass of soil solids (**lime limited**)	**0.280**	**0.560**	
Total mass of C-S-H + C-A-H produced by hydration and pozzolanic reactions/mass of soil solids, W_c	**0.432**	**0.696**	

of Equation (2.18) represents the amount of lime, which the hydration reaction of OPC can supply. As Table 2.5 shows, for binder content of 25%, the amount of lime available per unit mass of soil is 0.097. For binder content of 50%, the amount of lime available per unit mass of soil is 0.195, which just exceeds the mass of lime required to fully react with the silica and alumina in the soil. This suggests that, for the assumed parametric values, using an OPC content of slightly under 50% will generate just enough lime to fully react with the pozzolanic kaolinite, thereby creating a "balanced" pozzolanic reaction. Using more OPC than this will result in unreacted lime, which may offset part of the strength gain. Chew et al.'s (2004) results showed that strength continues to increase for OPC content higher than about 50%, because of strength gain due to continuation of hydration reaction. However, the rate of increase in strength, with respect to OPC content, moderates significantly when the latter exceeds about 50%. This may be explained by the slowing down or cessation of the pozzolanic reaction and presence of unreacted lime.

The mass of cementitious compounds produced per unit mass of untreated soil, W_c, can also be estimated using the following relationship:

$$W_c = A_w \psi_h + \psi_p \tag{2.19}$$

where ψ_h is the mass of cementitious compounds produced by the hydration reaction of unit mass of OPC and ψ_p is the mass of cementitious compounds produced by pozzolanic reaction with a unit mass of soil. ψ_p is limited by the amount of pozzolans available in the soil or the lime produced by the OPC. If the pozzolanic reaction is limited by availability of pozzolans, then

$$\psi_p = \psi_{p1} = \zeta_s P_s W_p + \zeta_a P_a W_p \tag{2.20}$$

where ζ_s and ζ_a are the amounts of C-S-H produced by unit mass of silica and C-A-H produced by unit mass of alumina, respectively. As previously discussed, ζ_s lies between 2.85 and 3.75 while ζ_a is about 4.4.

If the pozzolanic reaction is limited by the availability of lime, then

$$\psi_p = \psi_{p2} = \left\{\zeta_{ls}\varsigma + \zeta_{la}\left(1-\chi\right)\right\}\left(\lambda_2 P_2 + \lambda_3 P_3\right) A_w \tag{2.21}$$

where $\zeta_{ls} = \dfrac{\zeta_s}{\lambda_s}$ and $\zeta_{la} = \dfrac{\zeta_a}{\lambda_a}$ are the amounts of C-S-H and C-A-H produced by per unit mass of lime and χ is the proportion of lime which reacts with silica. Using the values of the parameters in Table 2.5 leads to ζ_{ls} ~ 2.357 and ζ_{la} ~ 3.5. There is some uncertainty over the proportion of lime which will react with silica χ versus that which will react with alumina $(1 - \chi)$. Table 2.5 shows that, based on Equation (2.20), if there is sufficient lime, the pozzolan in the soil will produce a mass of cementitious compounds equal to 0.39 times the mass of soil solids; this is a pozzolan-limited scenario. On the other hand, based on Equation (2.21), the available lime produced by the hydration reaction will produce a mass of cementitious compounds equal to 0.28 and 0.56 times the mass of soil solids, for 25% and 50% binder content, respectively. For cement content of 25%, the pozzolanic reaction is lime limited, that is $\psi_p = \psi_{p2}$, and the total mass of cementitious products W_c is approximately 0.432 mass-units per mass-unit of soil solids. For cement content of 50%, the pozzolanic reaction is pozzolan limited, that is $\psi_p = \psi_{p1}$, and W_c is approximately 0.696 mass-units per mass-unit of soil solids. These estimates are highly approximate since the values of many of the parameters are only estimates. Nonetheless, it may be useful for obtaining a first-order estimate for the mix proportions for pozzolanic reaction.

In addition, the framework permits us to explain some of strength gain features observed in cement stabilisation. The presence of unreacted lime in a pozzolan-limited scenario also explains the decrease in strength observed in Figure 2.1 for lime-treated sludge with lime content higher than 20%. The exhaustion of pozzolanic silica and alumina in the sludge leads to increase in the amount unreacted lime, which remains as a soft material in the soil matrix, thereby degrading its strength.

In some situations, pozzolans may be encapsulated by cementitious products to the extent that they are unable to react with available lime (e.g. Chew et al. 2004; Guo et al. 2022). Chew et al. (2004), which is attributable to the failure of illite to undergo pozzolanic reaction to the encapsulation of the clay particle surface by cementitious products. If this is so, then clay particle encapsulation also sets a limit to the extent of the pozzolanic reaction. Binder–soil reactions are generally complex. Nonetheless, by careful consideration of the stoichiometry of the hydration and pozzolanic reactions, some insights into the strength gain processes and mix balance may still be obtained.

REFERENCES

Bensted, J. (2008). Calcium aluminate cement. In *Structure and performance of cements* (Eds. P. Barnes and J. Bensted), 2nd Edition, Taylor & Francis.

Boura, P., Katsioti, M., Tsakiridis, P. and Katsiri, A. (2003). Stabilization/solidification of sewage sludge. *Proc. Int. Conf. BIOSOLIDS 2003 Waste-Water Sludge as a Resource*, Trondheim, Norway: 465–471.

Chen, H. and Q. Wang (2006). The behaviour of organic matter in the process of soft soil stabilization using cement. *Bulletin of Engineering Geology and the Environment*, 65(4): 445–448.

Chen, P., Feng, B., Lin, Y. and Lin, C. (2018). Solidification and stabilization of sewage sludge and MSWI bottom ash for beneficial use as construction materials. *Journal of Material in Civil Engineering*, 31(1): 04018351.

Chen, Y., He, X., Wang, P., Wan, Y. and Tan, X. (2020). Experimental study of moisture content effect on geotechnical properties of solidified municipal sludge. *Advances in Polymer Technology*, 2020: 8794076.

Chew, S. H., Kamruzzaman, A. H. M. and Lee, F. H. (2004). Physicochemical and engineering behavior of cement treated clays. *Journal of Geotechnical and Geoenvironmental Engineering*, 130(7): 696–706.

Davidson, L. K., Demirel, T. and Handy, R. L. (1965). Soil pulverization and lime migration in soil-lime stabilization. *Highway Research Record*, 92: 103–126.

Dung, N. T., Hooper, T. J. N. and Unluer, C. (2019). Accelerating the reaction kinetics and improving the performance of Na2CO3-activated GGBS mixes. *Cement and Concrete Research*, 126: 105927.

Fernandez, R., Martirena, F. and Scrivener, K. L. (2011). The origin of the pozzolanic activity of calcined clay minerals: A comparison between kaolinite, illite and montmorillonite. *Cement and Concrete Research*, 41(1): 113–122.

Guo, X., Zhang, W., Yu, H. and Jin, D. (2022). Reduction, stabilization, and solidification of Cr(VI) in contaminated soils with a sustainable by-product-based binder. *Chemosphere*, 307(2): 135902.

He, J., Zhang, L. and Zhang, C. (2021). Durability analysis of sludge solidified with soda residue subjected to dry-wet and freeze-thaw cycles. *Advances in Civil Engineering*, 2021: 9960416.

IUPAC (2021). *Atomic weights of the elements 2021*. https://iupac.qmul.ac.uk/AtWt/.

Kang, G., Tsuchida, T., Kim, Y. S. and Baek, W. (2017). Influence of humic acid on the strength behavior of cement-treated clay during various curing stages. *Journal of Materials in Civil Engineering*, 29(8): 04017057.

Katsioti, M., Giannikos, D., Tsakiridis, P. E. and Tsibouki, Z. (2009). Properties and hydration of blended cements with mineral alunite. *Construction and Building Materials*, 23(2): 1011–1021.

Katsioti, M., Katsiotis, N., Rouni, G., Bakirtzis, D. and Loizidou, M. (2008). The effect of bentonite/cement mortar for the stabilization/solidification of sewage sludge containing heavy metals. *Cement and Concrete Composites*, 30: 1013–1019.

Kulikowska, D., Klik, B., Gusiatin, M. and Hajdukiewicz, K. (2019). Characteristics of humic substances from municipal sewage sludge: A case study. *Desalination and Water Treatment*, 144: 57–64. https://doi.org/10.5004/dwt.2019.23622.

Lee, F. H., Lee, Y., Chew, S. H. and Yong, K. Y. (2005). Strength and modulus of marine clay-cement mixes. *Journal of Geotechnical and Geoenvironmental Engineering*, 131(2): 178–186.

Leonard, R. J. (1958). *The pozzolanic activity of certain fly ashes and soil minerals*. PhD dissertation, Iowa State University: 144p.

Li, H., Li, Y., Zou, S. and Li, C. (2014). Extracting humic acids from digested sludge by alkaline treatment and ultrafiltration. *Journal Materials Cycles and Waste Management*, 16: 93–100.

Lin, C., Zhu, W. and Han, J. (2009). Geotechnical properties of solidified sludge by mixing cement and calcium-bentonite. *Proc. International Foundation Congress and Equipment Expo—Contemporary Topics in Ground Modification, Problem Soils, and Geo-Support* (Eds. M. Iskander, D. F Laefer, and M. H Hussein), March 15–19, 2009, Orlando, Florida, ASCE: 281–288. https://doi.org/10.1061/9780784410233, ISBN (print): 9780784410233.

Lindgreen, H., Jacobsen, T. H. and Jakobsen, H. J. (1991). Diagenetic structural transformations in North Sea Jurassic illite/smectite. *Clays and Clay Minerals*, 39(1): 54–69.

Lucena, L. C. F. L., Juca, J. F. T., Soares, J. B. and Portela, M. G. (2014). Potential uses of sewage sludge in highway construction. *Journal of Materials in Civil Engineering*, 26(9): 1.

Malliou, O., Katsioti, M., Georgiadis, A. and Katsiri, A. (2007). Properties of stabilized/solidified admixtures of cement and sewage sludge. *Cement and Concrete Composites*, 29: 55–61.

Marshall, C. E. (1964). *The physical chemistry and mineralogy of soils*, Vol. 1: Soil Minerals, Wiley, New York.

Miller, G. A. and Zaman, M. (2000). Field and laboratory evaluation of cement kiln dust as a soil stabilizer. *Transportation Research Record*, 1714(1): 25–32.

Morse, J. W., Arvidson, R. S. and Lüttge, A. (2007). Calcium carbonate formation and dissolution. *Chemical Reviews*, 107(2): 342–381.

Neville, A. M. (1995). *Properties of concrete*, Vol. 4, Longman, London.

Palomo, A., Grutzeck, M. W. and Blanco, M. T. (1999). Alkali-activated fly ashes a cement for the future. *Cement and Concrete Research*, 29: 1323–1329.

Pertusatti, J. and Prado, A. G. S. (2007). Buffer capacity of humic acid: Thermodynamic approach. *Journal of Colloid and Interface Science*, 314: 484–489.

Rafalko, S. D., Filz, G. M., Brandon, T. L. and Mitchell, J. K. (2007). Rapid chemical stabilization of soft clay soils. *Transportation Research Records*, 2026(1): 39–46.

Salahudeen, N. and Idris, F. (2022). Effects of beneficiation on the characteristics of Alkaleri Kaolin. *Engineering Journal Chiang Mai University*, 29(1): 12–20.

Tremblay, H., Duchesne, J., Locat, J. and Leroueil, S. (2002). Influence of the nature of organic compounds on fine soil stabilization with cement, *Canadian Geotechnical Journal*, 39: 535–546.

Uddin, K., Balasubramaniam, A. S. and Bergado, D. T. (1997). Engineering behavior of cement-treated Bangkok soft clay. *Geotechnical Engineering*, 28(1): 89–119.

Weaver, C. E. and Pollard, L. D. (1973). *The chemistry of clay minerals*, Developments in Sedimentology, Vol. 15, Elsevier, Amsterdam: 213.

Wilkomirski, B. and Malawska, M. (2004). Characteristics of humic substances in peat of selected peatlands from North-Eastern Poland. *Soil Science and Plant Nutrition*, 50(6): 931–934.

Xue, Q. and Chen, Y. (2015). Experimental study on municipal sludge dewatering capacity by using quicklime and slag. *Desalination and Water Treatment*, 54(6): 1499–1506. https://doi.org/10.1080/19443994.2014.889607.

Zhu, X., Li, F., Zhang, J. and Wu, S. (2018). *Physicochemical properties of stabilized sewage sludge admixtures by modified steel slag*, De Gruyter: 16.

3 Mixing Quality

Fook-Hou Lee, Linqiao He, and Hui-Juen Teo

3.1 BACKGROUND

The goal of binder–soil mixing processes is often to produce a treated mass which is uniform in terms of strength and stiffness. In general, however, chemically stabilised soil mass invariably has a certain degree of non-uniformity. The parameters which are typically used to quantify the departure from a perfectly uniform mix are the coefficient of variation and the scale of fluctuation. Such non-uniformity typically arises from two sources. Firstly, the in situ soil may be non-uniform. Spatial variation in soil strength has been widely studied (e.g. Phoon and Kulhawy 1999; Kim 2005). However, Liu et al. (2019) noted that the strength of in situ soil is not readily correlated to the stiffness and strength of the soil after treatment, since the soil would have been remoulded and admixed with binders. Instead, they proposed that a more appropriate parameter for characterising spatial variation within a layer of deposited soft clay to be improved is the in situ water content. For instance, Liu et al.'s (2019) data from the Marina Bay Financial Centre site appears to show higher water content in the mid-layer of the soft clay, which is consistent with the fact that the soil is still consolidating. This is also correlated to the variation in strength of the cement-treated clay.

The second factor is the quality of mixing. In general, binder–soil mixing is invariably imperfect, in that there will always be some degree of heterogeneity in the admixture. The mixing quality is affected by numerous factors, including miscibility of the binder with soil, tendency of the components to segregate, mixing equipment as well as mixing effort.

3.2 IN SITU VARIATION IN SOIL PROPERTIES

In situ spatial variation in soil properties is often characterised in terms of its undrained shear strength or related parameters such as SPT blow counts and CPT resistance (e.g. Phoon and Kulhawy 1999; Kim 2005; Liu and Leung 2018). However, since mixing involves breaking up and remoulding of the soil matrix, as well as mixing with binder, the strength of the treated soil is not readily correlated to that of the in situ soil. State variables such as over-consolidation ratio and undrained shear strength are also unlikely to have a significant influence on the strength of improved soil, since the cutting, remoulding and mixing processes are likely to reduce, if not completely eliminate, the influence of over-consolidation ratio.

Other properties include those related to soil type and mineralogy, such as Atterberg limits, as well as the water content. Data on spatial variation in mineralogy

DOI: 10.1201/9781003285885-3

TABLE 3.1
Variation in in situ properties of some soft alluvial cohesive soils.

Location	Depth (m)	Undrained shear strength, c_u (kPa)	Natural water content[1] (%)	Liquid limit[1] (%)	Plasticity index[1] (%)	Data source
Hiroshima Port	9.0–22.0	3.5+1.5z (z = 0 at 9.0m depth)	110–150 [15]	125–140 [5.7]	80–110 [16]	Honjo (1982)
Yokohama Port	14.0–22.0	2.9z (z = 0 at 11.0m depth	40–100 [43]	70–85 [9.7]	40–50 [11]	
Tokyo Bay	9.0–39.0	3.0+1.4z (z = 0 at 9.0m depth)	100–130 [13]	80–120 [20]	20–40 [33]	
Bangkok, Sutthisan Station	4.0–15.0	5.0+2.7z (z = 0 at 0m depth)	~47.0–~75.0 [23]	~45.0–~99.0 [31]	~18.0–~26.0 [18]	Shibuya et al. (2001)
Bangkok, AIT	3.0–4.0	15.0+5.0z (z = 0 at 3.0m depth)	~80.0–~83.0 [1.8]	~90.0–~93.0 [1.6]	~60	Horpibulsuk et al. (2007)
Bangkok, Bangpee	3.0–15.0	15.0+1.7z (z = 0 at 0m depth)	~85.0–~110.0 [13]	~85.0–~110.0 [13]	~40.0–~80.0 [33]	
Singapore, Changi (Upper marine clay)	~5.0–~25.0	10.0 + 2.2z (z = 0 at 0m depth)	70.0–88.0 [11]	50.0–82.0 [24]	30.0–62.0 [35]	Bo et al. (2015)
Singapore, Changi (Lower marine clay)	~5.0–~25.0	~35.0–~51.0	40.0–60.0 [20]	63.0–83.0 [14]	43.0–62.0 [18]	

[1] Numbers in square brackets [] represent the half-range normalised by the median value (in %).

within a single soil formation remain scarce to date. Reatto et al. (2008) presented data on variation in kaolinite and gibbsite content in near-surface soil down to a depth of about 0.8m in a sloping landform measuring 100m by 200m. However, the exploration covers only the near-surface zone and the result is likely to be strongly influenced by vegetation and weathering. Thus, it may not reflect mineralogical variation at larger depths. Xia et al. (2019) reported variation in Atterberg limits in residual soils in south China, which shows plasticity index variation of less than ±1.8%. On the other hand, as Table 3.1 shows, variations in liquid limit and plasticity index of soft alluvial cohesive soils are typically much larger. There are some indications that the Atterberg limits may influence the strength gain characteristics of soft marine clay. In particular, the influence of binder content appears to increase with the liquid limit and plasticity index. Nonetheless, the effect of Atterberg limits on the strength of cement-admixed soil is still not fully understood.

As Table 3.1 shows, there is also significant variation in in situ water content in soft alluvial deposits. In a thick layer of soft clay, water content is expected to decrease with depth below ground surface as the effective stress increases. However, this is not always the case. Ohtsubo et al.'s (2000), Shibuya et al.'s (2001) as well as Horpibulsuk et al.'s (2007) AIT data seem to show a reasonably well-defined trend of decrease with depth. On the other hand, Bo et al.'s (2015) data on Singapore Upper marine clay at Changi site show an abrupt drop from about 70% to about 30% from depth of 10m to about 13m, before increasing back to about 50% below that. Similarly, using data from the Marina Bay Financial Centre (MBFC) and Marina One (M1) sites, Liu et al. (2019) noted that the water content within the mid-layer of the Singapore upper marine clay appears to be higher than that of the upper and lower boundaries, although a decreasing trend with depth is still present when the water content is averaged within layers of 2m thickness. They attributed this to the fact that the marine clay is still consolidating under the reclamation sand fill. Hence, water content can vary substantially, and this variation is significantly influenced by changes in depositional and consolidation conditions.

The information provided by these studies relate mainly to the ranges of variation in soil properties. However, much less is known about their statistical characteristics, such as coefficient of variation (CV), and spatial structure, such as the scale of fluctuation (SoF). The CV is defined as the standard deviation normalised by the mean, that is

$$\mathrm{CV} = \frac{S_d}{\bar{x}} \tag{3.1}$$

where S_d is the standard deviation of the population (or sample) and $\bar{x}$ is the mean of the population (or sample). SoF describes the distance over which a parameter of a soil or rock are correlated. It is typically evaluated as a constant in the autocorrelation function used to fit the coefficient of correlation. A more detailed coverage on the SoF will be given in Chapter 5.

Spatial variation in water content in soil has been widely studied (e.g. Huisman et al. 2002; Brocca et al. 2007; Zarlenga et al. 2018). However, the majority of these studies relate to hydrological, agronomical and soil science interests and deal mainly with horizontal variation of near-surface water content. Studies relating to water content variation of deeper soil layers are much scarcer. Akbas and Kulhawy (2010) reported vertical SoF of in situ water content for Ankara clay ranging from 2.5m to 5.5m. Valdez-Llamas et al. (2003) reported horizontal and vertical SoFs for superficial soft clay in the Gulf of Mexico of 22.2m and 0.8m to 2.0m, respectively. For deeper deposits, they also reported horizontal and vertical SoFs of 1000m and 21m, respectively. Using water content data from the MBFC and M1 sites, Liu et al. (2019) estimated that the coefficient of variation in in situ water content of the upper marine clay layer to lie between 0.11 and 0.14, while that for the lower marine clay is about 0.06. The horizontal and vertical SoFs are 100m to 200m and 6m, respectively, for both layers of marine clay. More data on SoFs of in situ water content of natural and improved soils are presented in Table 5.4. In general, the horizontal SoF of soils is much larger than the vertical SoF. It is also possible that part of the vertical SoF may actually be a vertical trend. In principle, the SoF is only calculated from the random variation after detrending. However, the vertical trend itself may be difficult to determine owing to the complex depositional and consolidation conditions. Hence, it may be difficult to differentiate random from deterministic variations.

As will be discussed in Chapter 4, the water content of a chemically stabilised soil is related to the void ratio, which has an important effect on its unconfined compressive strength. The latter typically decreases as the water content increases. Liu et al. (2019) postulated that the in situ spatial variation in water content is correlated to the strength of improved soil since higher water content of the in situ soil will also result in higher water content, and therefore higher void ratio, of the improved soil. In columnar mixing, the horizontal SoF in binder concentration is limited by the diameter of the deep-mixed column. On the other hand, horizontal SoF in in situ water content is much larger. The interaction of these two factors can give rise to two levels of horizontal spatial variation in the strength of the improved soil. The first is a short-range SoF caused by non-uniform binder distribution. The second is a long-range SoF arising from in situ water content variation that is much larger than column diameter. The existence of these two very different SoFs has implications on quality control and mass performance of the improved soil. In core sampling for quality control of deep-mixing works, close sampling of less than a column diameter may be needed to shed light on mixing-induced spatial variation in strength. On the other hand, the long-range SoF may significantly impact the overall performance of the improved soil mass, especially when strength and stiffness in the horizontal direction is required. An example of this is in the use of chemically improved soil for deep excavation support. This will be further discussed in Chapter 5.

3.3 MISCIBILITY

The term "miscibility" characterises the ease with which soil and binder can mix with each other to a required degree of uniformity. Several steps are required before soil and binder can be successfully mixed. The soil mass must first be cut and its structure broken up. Then binder fluid or water must be entrained with the soil debris in order to liquify the latter. Finally, the soil slurry must be mixed with the binder, which may be in powder or liquid form. The initial cutting of the soil mass is often accomplished using a continuous or segmental screw auger blade near the end of the rotary drilling shaft (Figure 1.13b). This also helps to facilitate penetration of the drilling equipment into the ground. The subsequent pulverisation of the cut soil debris and mixing with the binder is typically achieved through a series of mixing blades mounted further up the rotary shaft.

Soil mixing with binder in slurry form involves liquid–liquid mixing. Slurry binder is used if the water content of the in situ soil is too low to permit effective mixing. This is usually the case if the in situ water content is lower than the liquid limit of the soil. Depending upon the viscosity of the liquids involved, liquid–liquid mixing can occur in the laminar or turbulent flow regimes. The dimensionless number which determines the flow regime is the Reynolds number (Re), which expresses the ratio between inertial and viscous forces in the fluid. The Reynolds number is typically defined as

$$\mathrm{Re} = \frac{\rho v L}{\mu} \tag{3.2}$$

where ρ is the density of the fluid, v a characteristic velocity, L a characteristic length and μ the dynamic viscosity of the fluid. In general, if Re < 2000, then the flow is laminar and if Re > 4000 then the flow is fully turbulent. For 2000 < Re < 4000, the flow is in a laminar–turbulent transition regime. Equation (3.2) applies only to Newtonian fluids, wherein the dynamic viscosity, which expresses the ratio between shear stress and shear strain rate, is a constant. However, numerous studies have shown that soil slurry is non-Newtonian. Khatib and Richardson (1984) noted that the flow of kaolin slurry in pipes can be characterised using a power-law relationship between shear stress τ and shear strain rate $\dot{\gamma}$ of the form

$$\tau = K\dot{\gamma}^n \tag{3.3}$$

where K is the consistency index with unit of Pa.s^n and n is the dimensionless flow behaviour index. Figures 3.1(a) and (b) show Yi et al.'s (2018) parallel plate viscometer measurements for kaolin with water contents of 72% and 87%, respectively. The liquid limit of the kaolin is 80%; hence, these water contents represent 0.9 and 1.09 times the liquid limit, respectively.

Yi et al. (2018) fitted the data using Herschel and Bulkley's (1926) relationship of the form

$$\tau = \tau_y + K_1\dot{\varepsilon}_s{}^n \tag{3.4}$$

where τ_y is the yield stress at very low strain rate, $\dot{\varepsilon}_s$ is the generalised shear strain rate and K_1 is a fitted parameter. The generalised shear strain ε_s is a strain quantity which reflects the contribution of all the strain components to shearing and is defined by the following equation:

$$\varepsilon_s = \frac{2}{3}\sqrt{\frac{1}{2}\left\{\left(\varepsilon_x - \varepsilon_y\right)^2 + \left(\varepsilon_y - \varepsilon_z\right)^2 + \left(\varepsilon_z - \varepsilon_x\right)^2\right\} + \frac{3}{4}\left(\gamma_{xy}^2 + \gamma_{yz}^2 + \gamma_{zx}^2\right)} \tag{3.5}$$

where ε_x, ε_y and ε_z are the normal strains in the x-, y- and z-directions, respectively and γ_{xy}, γ_{yz} and γ_{zx} are the engineering shear strain components in the x-y, y-z and z-x planes, respectively. Where the soil is subjected to pure shear in the x-y plane only, then only γ_{xy} is non-zero and $\varepsilon_s = \frac{\gamma_{xy}}{\sqrt{3}}$. Hence Equation (3.5) can be converted to the more common Herschel-Bulkley form:

$$\tau = \tau_y + K\dot{\gamma}^n \tag{3.6}$$

where $K = \frac{K_1}{\sqrt{3}}$. The behaviour of cement slurry is also non-Newtonian.

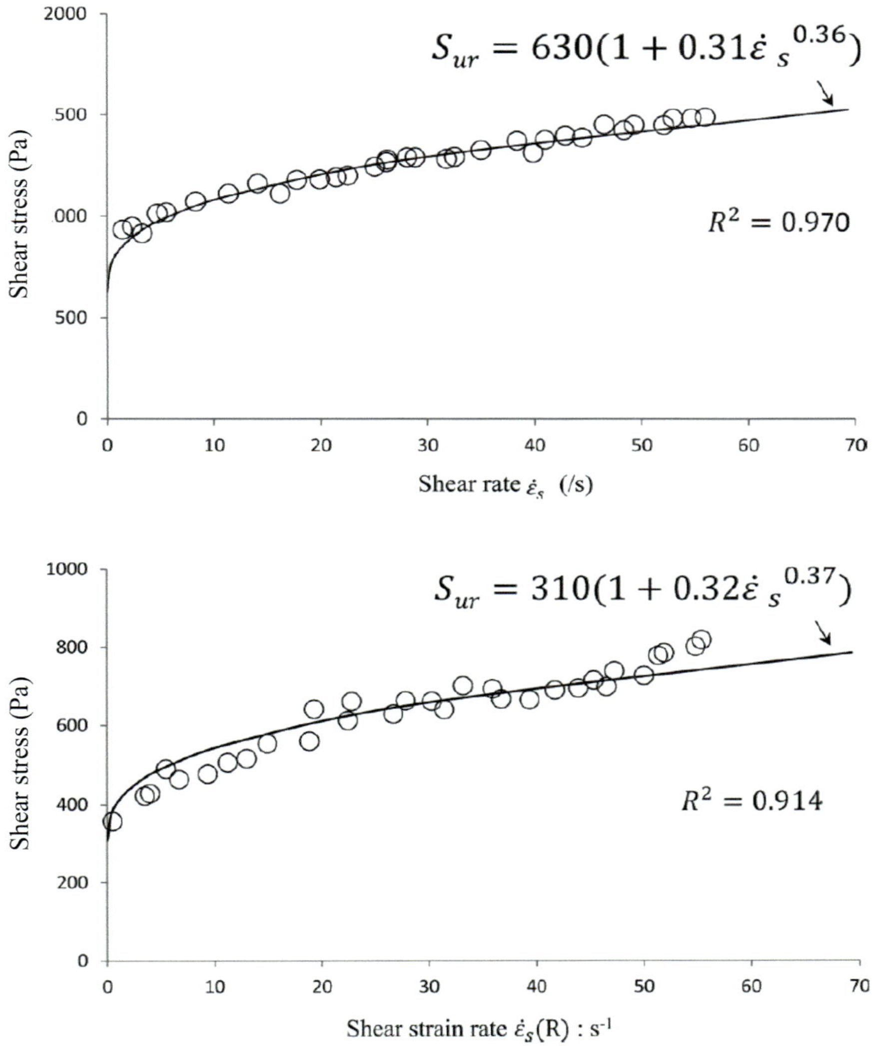

FIGURE 3.1 Shear stress versus generalised shear strain rate for kaolin at (a) 72% and (b) 87% water content.

(*Source:* Yi et al. 2018).

Figure 3.2 shows the shear stress versus shear strain rate for cement slurry, re-plotted using Lee et al.'s (2006) viscometer measurements. As can be seen, for similar water content, the Bingham yield stress τ_y and the consistency index K of cement slurry is much lower than that of kaolin. As Figure 3.3 shows, for shear strain rate below 100rad/s, the shear stress versus shear strain rate behaviour of the kaolin–cement admixture at different water and cement contents for strain rates up is better represented by a

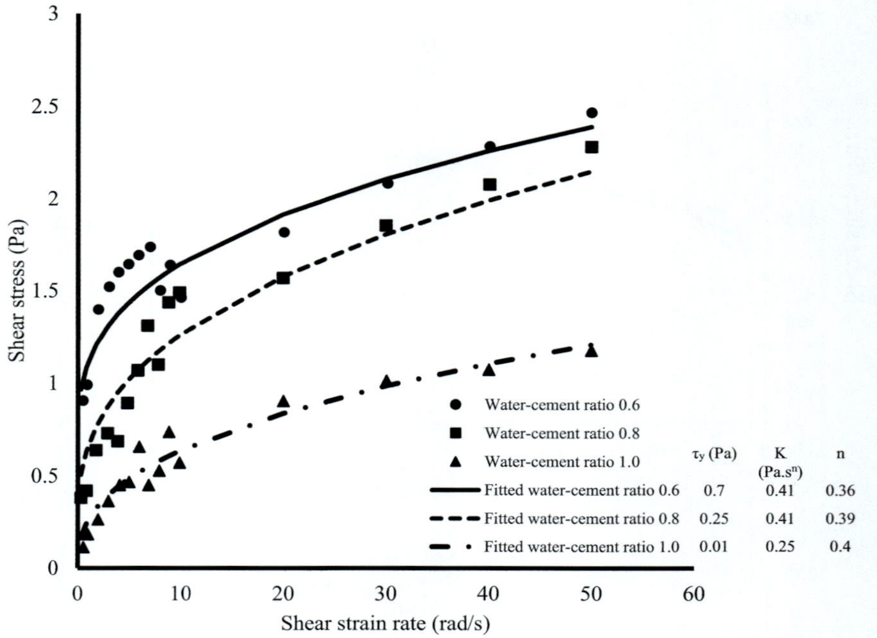

FIGURE 3.2 Shear stress versus shear strain rate of cement slurry for different water–cement ratios. Markers show measured data while lines show fitted relationships.

(*Data source:* Lee et al. 2006).

linear relationship (i.e. $n = 1$), with a Bingham yield stress that lies between that of pure kaolin and pure cement slurry.

The Reynolds number can be estimated from the aforementioned data using Lee et al.'s (2006) approach, which expresses the Reynolds number directly as a ratio of inertial forces to shear forces. This leads to

$$\mathrm{Re} = \frac{\rho v^2}{\tau} \tag{3.7}$$

The flow pattern within the deep-mixing cavity is still not well understood. For multi-shaft mixing equipment, wherein successive layers of mixing blades rotate in opposite directions, the average shear strain rate $\dot{\gamma}_{\mathrm{ave}}$ is likely to be related to the relative flow velocity Δv at the blade levels divided by the inter-blade spacing S between two adjacent shafts (Figure 3.4). The relative flow velocity Δv may, in turn, be characterised in terms of the tip velocity of the blades. This leads to the expression

$$\dot{\gamma}_{\mathrm{ave}} \propto \frac{2\pi Rr}{S} \tag{3.8}$$

where R is the rate of rotation in revs/s and r the blade radius. Lee et al. (2006) proposed a definition for the average shear strain rate $\dot{\gamma}_{\mathrm{ave}}$ between successive mixing blades given by

$$\dot{\gamma}_{\mathrm{ave}} = \frac{8\pi Rr}{3S} \tag{3.9}$$

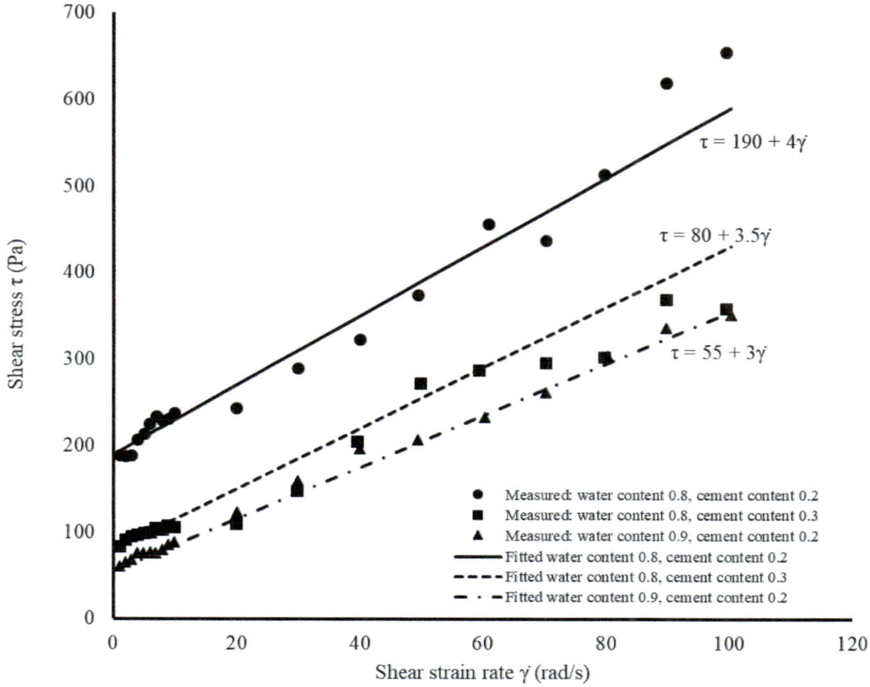

FIGURE 3.3 Shear stress versus shear strain rate of different cement–soil admixtures with different water and cement contents. Markers show measured data while lines show fitted relationships. (*Data source:* **Lee et al. 2006).**

(a) (b)

FIGURE 3.4 (a) Multi-shaft deep mixing rig and (b) idealised motion of two adjacent blade layers (belonging to two adjacent shaft axes) in the common plane of the shaft axes.

Source: **Courtesy of Presscrete Engineering Pte. Ltd.**

Table 3.2 shows the typical ranges of equipment and operating parameters in deep soil mixing. Table 3.3 shows the Reynolds number calculations using the parameters in Table 3.2. Mixing blades with a radius of 0.5m are considered. The rate of rotation was assumed to be 60RPM, which is the upper limit shown in Table 3.2. The kaolin is assumed to have a water content of 87%, which is higher than the typical liquid limit of kaolin. The cement slurry is assumed to have a water–cement ratio of 1.0 while kaolin–cement mixture has water content of 90% and cement content of 20%. These values are fairly typical of those encountered in deep-mixing operation.

Equation (3.9) is used to estimate the average shear strain rate shown in Table 3.3. The results show that the Reynolds number for kaolin and kaolin–cement admixture lies well below the laminar–turbulent transition level of 2000. The Reynolds number for cement slurry lies close to the laminar–turbulent transition range of 2000 to 4000, but is not high enough for full turbulent flow. At lower RPM, the Reynolds number will decrease significantly since inertial stress decreases as the square of the flow velocity (and thus RPM), whereas shear stress decreases only as a fractional power of the RPM; see Equation (3.6). For other soil types, similar calculations can be performed using measured viscosity data.

These results imply that, within the deep-mixing cavity, the flow regime is likely to be largely laminar. At the initial stages of mixing, transient turbulent flow may be present at locations with unmixed cement slurry. However, as mixing of cement and soil progresses, the viscosity of the cement–soil slurry should increase as the soil content increases, thereby decreasing the Reynolds number and bringing the flow into laminar regime. Laminar mixing occurs through the transfer of material across the shearing layers of fluid. On the other hand, in turbulent mixing, the fluid no longer travels in layers and mixing can occur through local eddies. In general, turbulent mixing is more rapid than laminar mixing. In both types of mixing, increasing the fluid viscosity will typically decrease the rate of mixing. Hence, decreasing the

TABLE 3.2

Typical parameters for deep soil mixing equipment.

Parameter	
Dimensions	**Typical values or ranges**
Maximum depth	About 40m
Diameter	0.6m–1.5m
	RAS jet equipment can produce column diameters up to 2.5m, which is considerably larger than the blade diameter.
	Operating parameters
RPM	12–60 (typical range)
Torque	15kN-m–30kN-m
Power	120kW–250kW
	Mix ratio
Binder content	10%–50%
Water–cement ratio (for cement binders)	90%–110%

TABLE 3.3

Calculation of flow regime for different operating parameters.

	Scenarios		
Parameters	**A**	**B**	**C**
	Geometry		
Radius (m)	0.50	0.50	0.50
RPM	60.00	60.00	60.00
Blade spacing s (m)	0.50	0.50	0.50
Strain rate $\dot{\gamma}$ (/s)	502.65	502.65	502.65
	Viscous shear stress		
Material	Kaolin	Cement	Kaolin–cement mixture
Water content or water–cement ratio	0.87	1.00	0.90
Cement content	0.00	1.00	0.20
Viscous shear stress τ (Pa)	1118.61[1]	7.50[2]	700.00[2]
	Inertial stress		
Density (kg/m^2)	1400.00	1500.00	1450.00
Flow velocity (m/s)	3.14	3.14	3.14
Inertial stress, ρv^2 (Pa)	13817.45	14804.41	14310.93
Reynolds number, Re	12.35	1973.92	20.44

[1] Estimated from fitted relationship in Figure 3.1b.

[2] Estimated from Lee et al.'s (2006) data.

water content of the admixture will typically increase the viscosity of the admixture, and thereby lead to a decrease in mixing efficiency. In particular, if the water content is too low, the Bingham yield stress of the admixture becomes so high that it adheres to the mixing blade, co-rotating with the latter. In such situations, circulation of the admixture ceases and mixing becomes impossible. This typically occurs when the water content of the admixture is lower than the liquid limit of the soil. Model tests in the laboratory indicate that a good range of water content for mixing is between 1.3 to 1.5 times the liquid limit. Hence, there is usually a lower limit in the water content that is required for mixing. For this reason, water is commonly added to the soil, usually as part of the binder, during mixing. It is thus not unusual for the admixed soil to have a higher water content, and thus void ratio, than the in situ soil.

Another approach is to use binder in dry powder form. This approach may be used when the in situ water content is substantially higher than the liquid limit of the soil. An example is mixing lime with dredged clay or sludge with high water content. Another instance is dry lime or cement mixing with Scandinavian quick clays, which have in situ water content higher than their liquid limit. In such cases, the binder is dispersed into soil matrix to form a slurry. Nonetheless, the subsequent homogenisation process is likely to be similar to that of liquid–liquid mixing.

3.4 SEGREGATION

While too low a water content renders the binder–soil admixture immiscible, too high a water content may also not be conducive to good mixing. Yoshizawa et al. (1997) reported a field-data survey on the factors affecting the strength and uniformity of deep-mixed soil columns, in which they noted that smaller spatial variation in strength was achieved by reducing the water–cement ratio of the binder, increasing the binder quantity and increasing the number of rotations of the mixing blade. Lee et al. (2006, 2008) attributed the effect of water–cement ratio to its effect on the density of the binder. Using centrifuge models of kaolin beds with soil density of 1630kg/m^3 and binder with densities of 1700kg/m^3, 1500kg/m^3 and 1300kg/m^3, they noted that as the binder density decreases from 1700kg/m^3 to 1300kg/m^3, the CV in binder concentration increases by nearly 100%, from about 0.2 to about 0.4 (Figure 3.5). In Figure 3.5, the density of the binder is expressed as a density ratio, which is defined as the ratio of binder density to soil density. Similar decreases in CV was also reported by He (2018) over a wide range of rotation speeds.

The specific gravity of OPC powder lies between 3.1 and 3.16. The binder densities of 1300kg/m^3, 1500kg/m^3 and 1700kg/m^3 correspond to a water–cement ratio of approximately 1.95, 1 and 0.65, respectively. Yoshizawa et al.'s (1997) strength CV data, in terms of water–cement ratio, was converted to slurry density, based on a specific gravity of cement of 3.15. Strictly speaking, Lee et al.'s (2006) results and Yoshizawa et al.'s (1997) results cannot be compared directly, since the former relates to binder concentration whereas the latter relates to unconfined compressive strength. Nonetheless, the similarity in trend is evident. Matsuo et al. (1996)

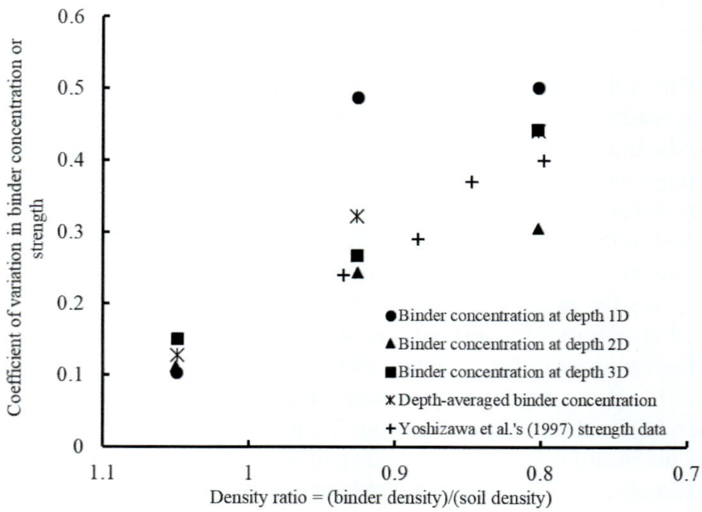

FIGURE 3.5 Change in CV in binder concentration with density ratio from Lee et al.'s (2006) centrifuge model test data and Yoshizawa et al.'s (1997) field-measured strength CV data. Blade rotation number used in centrifuge model tests is 800.

also reported a similar increase in strength CV as water–cement ratio increases and suggested that the degradation in mixing quality can be attributed to buoyancy effects due to density differences between slurry and soil. All these indicate that the increase in strength CV with water–cement ratio may be attributed, at least in part, to the effect of density differences.

The effect of density differences in liquid–liquid mixing is well recognised (e.g. Bouwmans et al. 1997; Rielly and Pandit 1988; Derksen 2011). The mixing behaviour of two liquids with different densities is usually characterised in terms of the Richardson number Ri, which is defined as the ratio of buoyancy forces to inertial forces, that is

$$Ri = \frac{\Delta \rho g L}{\rho v^2} \tag{3.10}$$

where $\Delta\rho$ is the density difference of the two liquids, L is a characteristic dimension, ρ is a reference density, typically taken to be that of one of the liquids, and v is a characteristic velocity. Bouwmans et al. (1997) studied the blending of liquids with different viscosities and densities in cylindrical containers with a 45° pitched, six-bladed impeller rotated so as to pump liquid downwards. They defined the Richardson number by

$$Ri = \frac{\Delta \rho g H}{\rho (RD)^2} \tag{3.11}$$

where H is the height of the container, R is the rate of rotation of the mixing tool in radians per second and D is the blade diameter. Their results showed that, at low Ri less than about 1, density differences are sufficiently small so that buoyancy forces, which keep the liquid segregated, can be overcome by large-scale circulatory flow. This flow regime was termed as the stirrer-controlled regime. For Ri greater than about 1, buoyancy forces are stronger and part or all of the lighter liquid tends to rise to the surface while the denser liquid sinks to the bottom. Subsequent mixing is then achieved by turbulent eddies, which lead to much longer mixing times. This was termed the gravity-controlled regime. Bouwmans et al.'s (1997) results show that the transition from stirrer- to gravity-controlled regimes occur at Ri between 0.1 and 1. Similar effects have also been noted by Burmester (1993). Bouwmans et al.'s (1997) results show that density differences as small as $100 kg/m^3$ are sufficient to bring about gravity-controlled mixing. Hence, small density differences between the liquids to be mixed can significantly influence the quality of mixing.

In the mixing of binders and clayey soil, the cut soil debris may initially be in a solid form and its density will be close to that of the in situ soil. As the debris becomes entrained by water and binder, its density may gradually converge to that of the admixture. To avoid segregation of soil debris prior to homogenisation, the binder slurry should have density close to that of the in situ soil. High water–cement ratio in cement slurries will lead to low binder density and may allow cut soil debris to sink to the bottom before homogenisation. A water–cement ratio of 1.0 will give a cement slurry density of about $1500 kg/m^3$, which may be slightly lighter than the density of soft clays, resulting in sub-optimal mixing condition. Using a water–cement ratio of about 0.9 will give a cement slurry density of about 1560kg/m^3, which may be closer to the density of soft clays. A water–cement ratio higher than 1.0 is likely to accentuate density difference and degrade mixing quality. The resulting heterogeneity will increase the strength CV of the treated soil, even though the segregation may not be apparent to the naked eye.

Where the in situ soil contains significant amounts of sand, gravel and even rock, segregation of a different form may arise, with the denser granular fraction sinking to the bottom. This can also potentially lead to top-to-bottom heterogeneity in the admixture, if the binder–clay mix is unable to keep the granular fraction in suspension. Hence water–cement ratio may also need to be kept sufficiently low in such situations. The increase in viscosity of cement slurry as its water–cement ratio decreases (Figure 3.2) may inhibit mixing slightly but it also helps to keep the heavier particles in suspension for longer durations and thus forestall segregation.

3.5 DEGREE OF MIXING

The degree of mixing is not synonymous to the quality of mixing. Instead, it refers to the effort, but not necessarily energy, which is invested into the mixing process. In general, the mixing effort is related to the following parameters:

1. The number of revolutions which the mixing tool has rotated through per unit volume of soil. Higher number of revolutions is considered to represent greater mixing effort.
2. The number of mixing blades on the mixing tool. Higher number of mixing blades is considered to confer greater mixing effort.
3. The design of the mixing tool and blades. This is also likely to influence the efficacy of mixing.
4. The characteristics of the soil and binders to be mixed, especially their shear stress–shear strain rate characteristics.

The most common indicator for the degree of mixing is the blade rotation number, which was proposed by Yoshizawa et al. (1997) and subsequently adopted by CDIT (2002) for deep columnar mixing. For deep columnar mixing, where the binder is injected during the penetration phase, known as penetration injection mixing, Yoshizawa et al. (1997) proposed that the blade rotation number T be defined as

$$T = \left(\sum M\right) \cdot \left(\frac{R_\mathrm{d}}{v_\mathrm{d}} + \frac{R_\mathrm{u}}{v_\mathrm{u}}\right) \tag{3.12}$$

where
ΣM is the total number of mixing blades,
R_d and R_u are the rates of rotation in RPM, during the penetration and withdrawal phases, respectively, and
v_d and v_u are the penetration and withdrawal velocities, respectively, in m/min.

The quantities $\frac{R_d}{v_d}$ and $\frac{R_u}{v_u}$ represent the number of revolutions per metre depth of the deep-mixing cavity which each blade would have made during the penetration and withdrawal phases, respectively. Hence, the blade rotation number is the sum of revolutions made by all the blades on the mixing tool per metre depth of the cavity during the mixing process.

There are also deep-mixing operations in which the binder is only injected during the withdrawal phase. For such withdrawal injection mixing, the term $\frac{R_d}{v_d}$ vanishes and thus

$$T = \left(\sum M\right) \cdot \frac{R_u}{v_u} \tag{3.13}$$

The blade rotation number is a kinematic quantity that accounts for the number of mixing blades and the number of revolutions per unit volume of soil but not the design of the mixing blades and the torque on the mixing shaft. It is not an energy quantity.

The blade rotation number is only applicable to columnar mixing. For other mixing processes, there is no well-accepted indicator for degree of mixing. Bellato et al. (2017) proposed a specific energetic index E_{CSM} for quantifying the degree of mixing in cutter soil mixing (CSM), which is given by

$$E_{CSM} = \frac{2\pi \left(T_d R_d t_d \phi_d + T_u R_u t_u\right)}{V_c} \tag{3.14}$$

where the subscripts d and u refer to the penetration and withdrawal phases, respectively; T_d and T_u are the average torques on the cutter wheels; R_d and R_u are the rotational speeds (in RPM) of the cutter wheels; t_d and t_u are the durations used to blend the soil and binder over 1m depth; V_c is the volume occupied by the mixing wheels and ϕ_d is a phase factor, which is used to differentiate the mixing effort that is used for situations when the binder is introduced during penetration or withdrawal phases. If the binder is introduced during the penetration (i.e. downward) phase, then mixing occurs during both penetration and withdrawal and $\phi_d = 1$. This is termed one-phase construction method (Bauer 2021). If the binder is introduced only during the withdrawal (i.e. upward) phase, then soil is only mixed with bentonite and water during the downward penetration phase, for which Bellato et al. (2017) proposed that $\phi_d = 0.5$. This is termed two-phase construction method.

The blade rotation number is a generalised definition of the mixing time which accounts for the rotational speed and number of mixing blades. Bouwmans et al. (1997) used the product of the rotational speed and time of mixing to compare the effect of density and viscosity density on mixing quality. This is similar to the blade rotation number except that the number of mixing blades is not considered. In contrast, E_{csm} is an energy quantity. The terms $T_d R_d t_d$ and $T_u R_u t_u$ represent the work done by the cutter wheels during the downward penetration and upward withdrawal phases. It should be noted that E_{CSM} will increase as the torque on the mixing wheels increases, say, as a result of stronger soil or more viscous admixture being encountered. What is unclear is how the resistance encountered by the mixing wheels contributes to mixing effort, even though it increases the energy input.

The effect of blade rotation number on columnar mixing quality is shown in Figure 3.6, which incorporates Chen et al.'s (2016), Lee et al.'s (2008) and Mizuno et al.'s (1988) results. Mizuno et al.'s (1988) results relate to coefficient of variation of unconfined compressive strength from core samples in the field while Chen et al.'s (2016) and Lee et al.'s (2008) results relate to coefficient of variation of binder concentration from centrifuge model tests. Both of these are indicators of the quality of mixing. Lee et al.'s (2008) data were obtained using three different mixing tool designs on normally consolidated kaolin clay beds while Chen et al.'s (2016) data were obtained from kaolin clay beds with over-consolidation ratios of 1.3, 1.5 and 1.7. Notwithstanding the different mixing tools and over-consolidation ratios, Lee et al.'s (2008) and Chen et al.'s (2016) are banded according to the binder density.

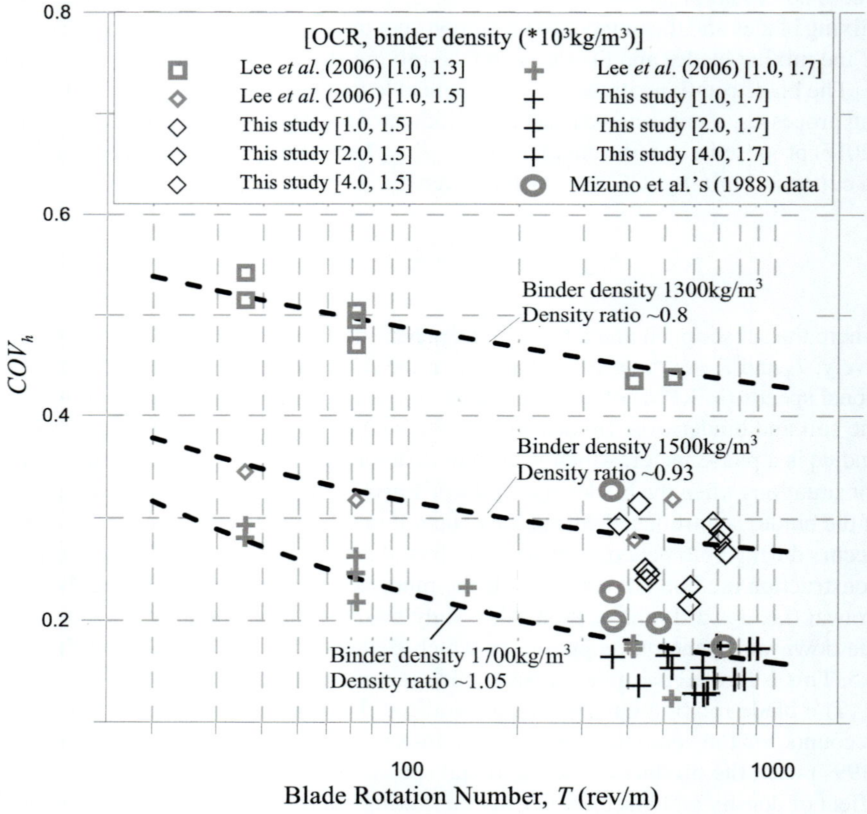

FIGURE 3.6 Coefficient of variation in binder concentration for different blade rotation numbers and binder densities.

Data source: **Lee et al. 2008; Chen et al. 2016.**

Mizuno et al.'s (1988) field data on coefficient of variation of unconfined compressive strength is also included. Density ratio is the ratio of binder density to a nominal density of 1630kg/m³ for soft kaolin clay.

Moreover, in all cases, the coefficient of variation decreases with blade rotation number, albeit at a diminishing rate. Based on Mizuno et al.'s (1988) data, Yoshizawa et al. (1997) propose a minimum blade rotation number of 360. However, Mizuno et al.'s (1988) data are relatively limited. Figure 3.6 shows that, at blade rotation number higher than about 400, the rate of decrease in the coefficient of variation moderates considerably. Hence, a blade rotation number of about 400 may be considered a minimum threshold for mixing quality to attain a stable level.

Teo (2017) investigated the effect of high blade rotation numbers up to 1000 on CV, using 1g $\frac{1}{3}$-scale models of deep-mixed columns of kaolin and cement slurry. Although her experiments were conducted in 1g, the low-scaling ratio should help to ensure that the model soil columns are reasonably representative of prototype soil columns. The mixing was also conducted using single-stacked, four-bladed mixing tools. Teo's (2017) data indicate that CV will continue to decrease as blade rotation number increases to 1000, albeit at a diminishing rate. For practical purposes, a blade rotation number of 400 can be considered to represent an acceptable level of mixing effort. However, for a given blade rotation number, that is mixing effort, the mixing quality is also dependent upon other factors such as density differences, which affect segregation tendency, RPM and equipment design. The influence of RPM and equipment design will be discussed further.

For shallow or mass soil mixing, there is no well-accepted indicator of the mixing effort or degree of mixing. One of the difficulties in defining an appropriate indicator is that, in contrast to deep mixing, there is no well-defined penetration-withdrawal sequence. In many instances, the cutter-mixer head (see Figure 1.19) is swiped in both horizontal and vertical directions. Hence, it is more difficult to define a control volume.

Figure 3.7 illustrates a possible approach to defining the mixing effort for the case of a horizontally aligned rotary cutter-mixer head, if the zone to be improved can be

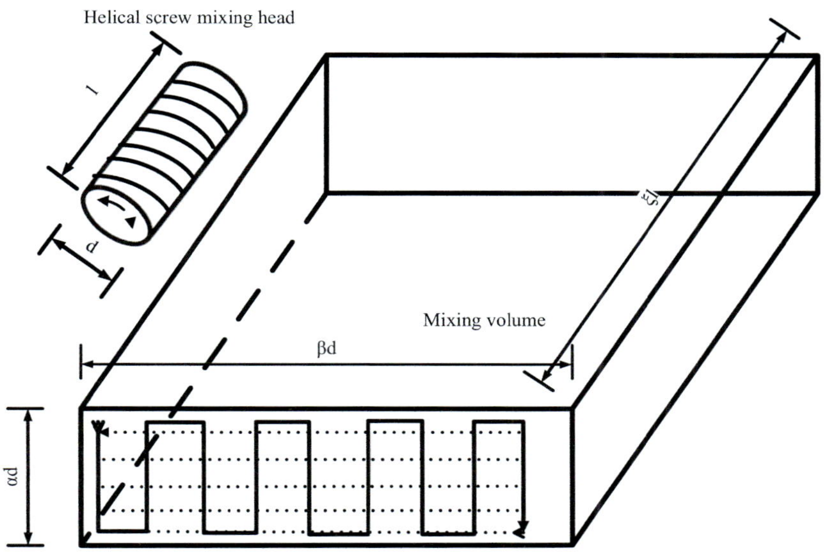

FIGURE 3.7 Cylindrical cutter-mixer head for shallow mass soil mixing with controlled mixing volume.

subdivided into regular shaped control volumes. In this figure, the cutter-mixer has an effective mixing diameter of d and length l and is rotated at a rotational speed R for time duration t. If a regular sweep sequence is considered, such as that which involves moving the cutter-mixer in an up–down direction, followed by horizontal sweeping motion, then the average mixing effort or degree of mixing M_e over the control volume can be approximately described by the following ratio:

$$M_e = \frac{\text{Effective mixing volume of cutter-mixer} \times \text{Number of revolutions}}{\text{Volume of soil}} = \frac{\pi}{4}\frac{Rt}{\alpha\beta\xi} \tag{3.15}$$

where α, β and ξ are the height, width and length of the control volume, expressed as multiples of diameter d and length l of the cutter-mixer head. Equation (3.15) does not consider the number of mixing fins on the cutter-mixer. Shallow mixing equipment tends to be more varied in design. Some designs have mixing fins; others feature largely continuous helical screws while some combine helical screws with fins mounted on the outer periphery. Hence, it is difficult to compare degree of mixing across different equipment.

3.6 ROTATIONAL SPEED

The effect of rotational speed in columnar mixing has not been widely studied. However, field observations suggest that the strength of soil-mixed column may be higher at the centre than the periphery. For deep-mixing works in Singapore, core sample specifications often stipulate that samples be taken at the column centre, at ⅔ radius from the centreline and at the column overlapping areas. For instance, LTA (2020) specifies that, for jet grouting trials, "the effectiveness of the trial grouting shall be demonstrated with a minimum of four boreholes used to obtain cores through the full depth of the grouted block. Of the four boreholes, one shall be targeted at the centre of a pile, one at a point two third of a pile radius from the centre, and the remaining two at the overlapping areas of the piles." For quality control of jet grouting, LTA (2020) also specifies that "there shall be 8 boreholes for each 1,000 cubic metres (rounded up to the nearest whole number of boreholes) of treated soil. The boreholes shall be at the overlapping area of the piles. In each borehole, the SPTs tests shall be done at the top, middle and bottom of the treated zone." Similarly, for quality control of deep soil mixing works, "Boreholes shall be targeted within the midway between the centre and outer surface of the column to obtain samples for laboratory testing." All these suggest that there is an awareness that strength of chemically improved soil may change with radial distance from the centreline.

He (2018) investigated the effect of rotational speed on binder distribution using centrifuge models. Figure 3.8 shows the change in binder concentration with radial distance from the centreline of the column, for different prototype rotational speeds for a single-stacked, four-bladed mixing tool (designated as Blade A herein), with injection nozzle at the base of the rotor shaft. This nozzle arrangement is typical of that in many columnar mixing equipment. The blade rotation number used in the

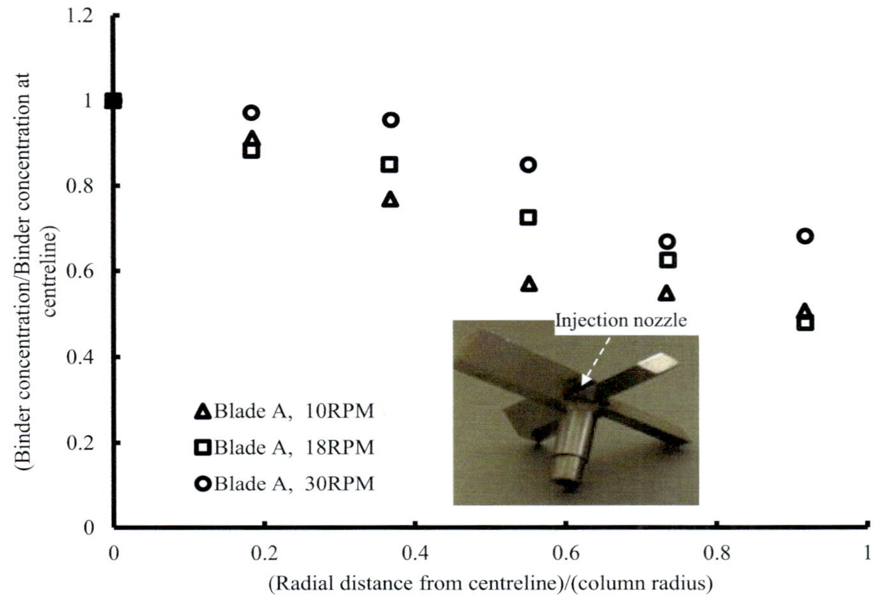

FIGURE 3.8 Variation of binder concentration with radial distance from centreline. Inset shows mixing blade design (designated Blade A) with location of injection nozzle. Binder density 1500kg/m^3.

Source: **He 2018.**

model tests was 360 revolutions per metre and the binder density was 1500kg/m^3. For each radial distance, the binder concentration is measured over 4 to 12 sampling points along the circumference and the average value was adopted. In general, the binder concentration decreases with radial distance from the centreline of the column for prototype-equivalent rotational speeds of 10RPM and 18RPM. However, the rate of decrease with radial distance moderates as the rotational speed (RPM) increases. At 30RPM, there seems to be some accumulation of binder at the column edge.

He's (2018) results indicate that the rotation of the mixing blades is not an efficient mechanism for ensuring radial distribution of the binder since it only induces circumferential flow; radial distribution is induced only indirectly by the centrifugal forces on the admixture. This is also evident from Teo's (2017) 1-g model observations on $\frac{1}{3}$-scale deep-mixed columns of kaolin and cement slurry. Although Teo's (2017) experiments were performed on 1g models, the low-scaling ratio should help to ensure that the model soil columns are reasonably representative of prototype soil columns. The mixing was also conducted using single-stacked, four-bladed mixing tools. Figure 3.9 shows the contours of cement concentration, with dark areas representing zones of higher cement concentration. These contours were obtained by digital processing of photographs of slices of the model clay bed after curing. As can be seen, the spread of the cement slurry is predominantly circumferential, indicating

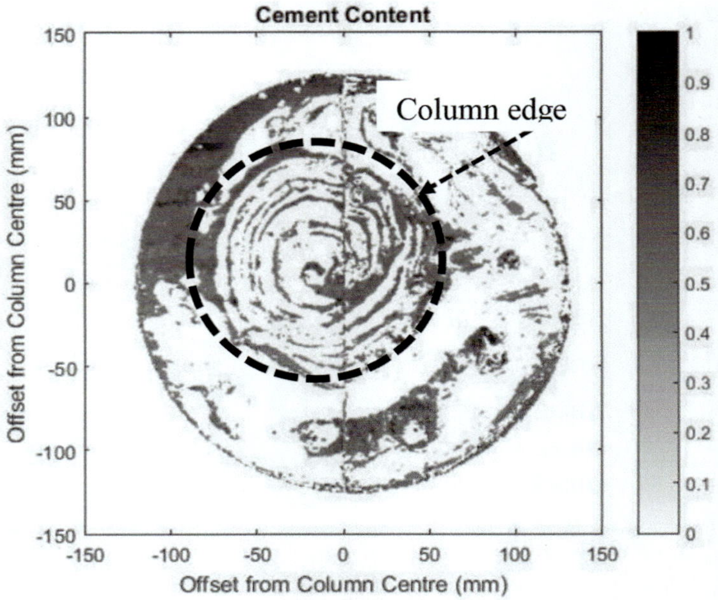

FIGURE 3.9 Contours of cement content in a hardened 1-g model deep-mixed column. Dark areas represent higher cement content.

Source: Teo 2017.

that mixing due to blade motion is also largely circumferential. The extent of radial mixing appears to be much less. Using higher rotational speed probably encourages better radial mixing due to the larger centrifugal forces which drive the soil–binder mix towards the column periphery. This sets up a higher pressure zone at the periphery, forcing inward radial flow of the mixture at inter-stack levels where the circumferential motion is slower. The outward flow of the soil–binder mixture at the level of the cutter blade stack and inward flow at inter-stack levels leads to circulation of soil–binder mixture, which helps to homogenise the admixture. However, significant radial heterogeneity still remains, as Figure 3.8 shows, indicating that the influence of the centrifugal forces is relatively limited.

The influence of rotational speed is also evident from the CV (He 2018), as Figure 3.10 shows. He (2018) investigated the influence of rotational speed on CV for three blade designs, with different nozzle arrangements, which will be discussed further. In general, CV decreases with rotational speed, except for the highest rotational speed of 30RPM, where there appears to be an increase in CV for Blades A and B. In the case of Blade A, this increase in CV is attributable to an accumulation of binder at the column edge, which was noted earlier (Figure 3.8). The influence of rotational speed can therefore be related to its effect on radial mixing, as previously discussed. For cutter soil and mass soil mixing, the effect of rotational speed of the cutting and mixing wheels is still not well understood.

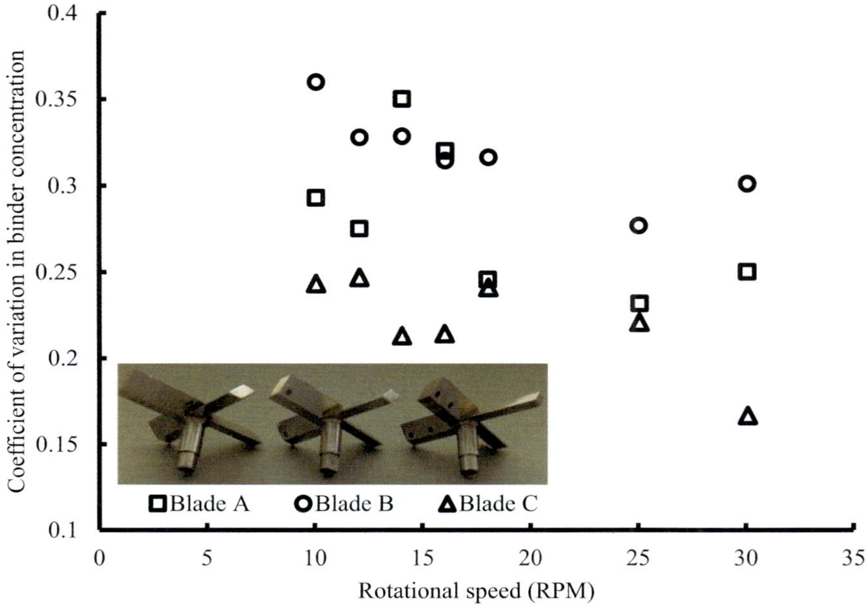

FIGURE 3.10 Variation in the coefficient of variation of binder concentration with rotational speed and nozzle arrangement. Binder density 1500kg/m^3. Blade rotation number 360 revolutions per metre.

Source: **He 2018.**

3.7 MIXING EQUIPMENT AND DESIGN

For columnar deep mixing, the mixing tool typically consists of a central rotor shaft with auger blade at the lower end of the shaft and multiple cutting-mixing blades further up along the shaft (Figure 1.13). A common variation of this basic configuration is multi-shaft mixing equipment, which can install a few columns in one penetration-withdrawal sequence. Research relating to the design of deep-mixing equipment is relatively scarce in the public domain. Amongst the recent works are those of Teo (2017) and He (2018). The latter investigated the influence of three nozzle location arrangements in a four-bladed mixing tool, as shown in Figure 3.10. Blade A has a single nozzle near the tip of rotor shaft. The second, designated Blade B, incorporates 4 additional nozzles, near the end of each mixing blade. The third, designated Blade C, incorporates another 4 additional nozzles, at the mid-span of each mixing blade. The blade rotation number used in all tests was 360 revolutions per metre.

As previously discussed, the main problem with Blade A is the accumulation of binder near the point of injection, that is the centreline. This effect can be mitigated by the use of higher rotational speeds. As Figure 3.10 shows, in terms of CV, the performance of Blade B was no better than that of Blade A. In fact, the decrease in CV with rotational speed is also lower for Blade B. In particular, at high rotational speed, Blade B gives higher CV

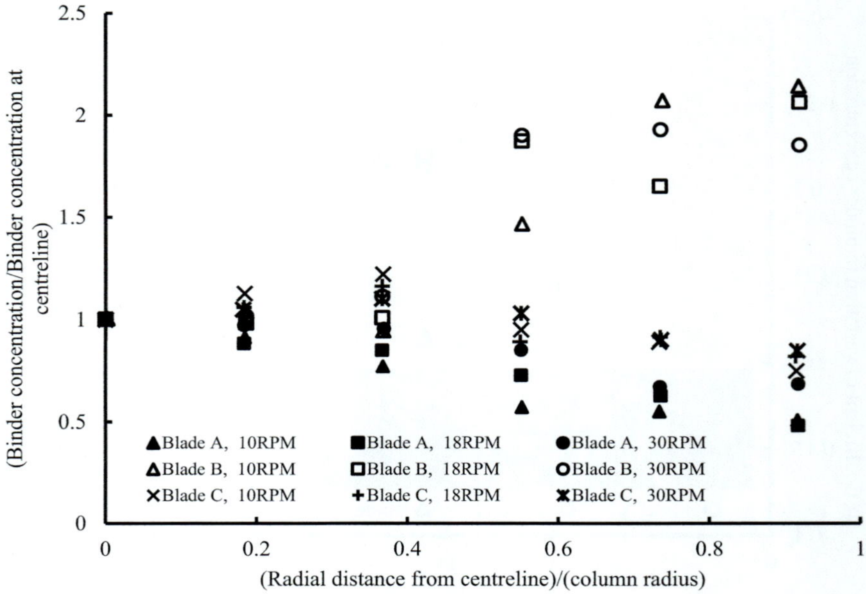

FIGURE 3.11 Variation of binder concentration with radial distance from centreline for blades A, B and C (see inset of Figure 3.10 for nozzle arrangements). Binder density 1500kg/m^3.
Data source: **He 2018.**

than Blade A. Figure 3.11 shows why this is so. The addition of four injection nozzles at the ends of the mixing blades leads to excessive binder injection at the periphery of the column. Increasing the rotation speed does not significantly mitigate this problem. Hence, the CV decreases less with rotational speed in Blade B, compared to Blade A.

Blade C gives the lowest CV of the three nozzle arrangements, over the range of rotational speed investigated, with the CV decreasing with rotational speed. As Figure 3.11 shows, Blade C gives the most uniform radial distribution in binder concentration amongst the three blades. All these indicate that the CV in binder concentration is significantly influenced by the radial distribution of binder, and current columnar mixing equipment may not be optimised for radial mixing. This is not surprising since the flow induced by the blade rotation is predominantly circumferential while the up–down movement of the shaft ensures a certain degree of mixing at all depths. However, there is no mechanism to impel radial flow directly. One way to mitigate radial heterogeneity is to increase the rotational speed. However, this has to be carefully calibrated to avoid excessive accumulation of binder at the column edge at high rotational speed. Using higher rotational speed also increases the power demand on the motor. Another way is by using multiple injection nozzles along the length of the blades so as to improve the radial distribution of the binder upon injection. However, the locations of the injection nozzles have to be judiciously chosen for it to have a beneficial effect. A third approach is to incorporate fins onto the mixing blades, which will induce some degree of radial mixing. Indeed, inducing radial flow may encourage circumferential as well as radial circulation, which is an important mechanism in the blending of miscible fluids. However, this may increase the torque required.

Another possible method of avoiding grout concentration at the column centre is by combining jet grouting and mixing, such as the RAS-JET method (Raito 2007). This method allows grout to be jetted out from the tips of the mixing blades, thereby creating an outer jet-grouted shell with an inner deep-mixed central region. This is similar to introducing grout at the tips of the blades, except that jet-grouting pressure is much higher, thus potentially allowing a bigger cavity to be cut.

3.8 POSITIONING ERROR

Liu et al. (2015) used the term "positioning error" in columnar deep mixing to refer to the error in the position of the deep-mixed column. The degree of importance of positioning error depends upon the usage of the soil stabilisation works. For instance, if the deep-mixed columns are used as foundation elements, a small deviation from intended position may not have a significant effect on its performance. On the other hand, for seepage cut-off and excavation support, where contiguity between columns is an important consideration, positioning error may have an important effect on the performance of the treated soil mass.

Table 3.4 summarises some recommended and allowable maximum deviations from design positions of jet-grouted and soil mixed columns. As can be seen, the deviation can be characterised in terms of a start position (i.e. ground surface) positioning error and an off-verticality. For deep jet grouting and soil mixing, off-verticality is often more important than the start position error, since a small amount of off-verticality can lead to a large error in columnar position at the depth

TABLE 3.4

Some published guidelines for allowable positioning error in deep-mixing and jet-grouting works.

	Maximum permissible deviation from prescribed position and verticality		
Mixing method	Rig position	Off-verticality(compared by drilling depth)	References
Jet grouting	50mm	0.5%	Morey and Campo (1999)
	50mm	2% for depths up to 20m, different tolerances for greater depths.	BS-EN (2001)
	–	0.5% for depths up to 15m, 1% for depths up to 30m.	Stoel (2001)
	50mm	0.5%	Passlick and Doerendahl (2006)
	76.2mm (3ins.)	1%	Geo-Institute/ASCE (2016)
Deep mixing	–	1%	Ryan and Jasperse (1989)
	50mm	1%	Bahner and Naguib (1998)
	76.2mm (3ins.)	2%	Puppala et al. (2008)
	–	1%	Rutherford et al. (2005)
	–	1%–2%	Larsson (2005)
	50.8 mm	2%	Puppala et al. (2008)

of improvement. For example, a 1% off-verticality over a depth of 20m will result in a lateral deviation of 20cm from the design position, which is a significant fraction of the column diameter. This can lead to a substantial loss of contiguity between columns and contribute to spatial variation in properties of the improved soil mass.

The effect of positioning effect on the performance of the improved soil slab comprising improved soil columns has been studied by Liu et al. (2015) using random finite element analyses. Such improved soil slabs are commonly installed within a deep excavation to help provide lateral support to the retaining walls below formation level. Their results indicate that, for an improved soil slab comprising treated soil columns, the lateral compressive capacity of the soil slab will decrease significantly if the positioning error exceeds about 30% of the column radius. For a column with diameter of 1m, this will imply a limiting positioning error of 0.15m. For the case of an improved soil slab to be installed at 10m below ground surface, this is equivalent to a limiting off-verticality of approximately 1.5% or 0.86°. Hence, positioning error is a workmanship issue even though it is not directly related to quality of mixing.

3.9 SUMMARY OF FACTORS AND THEIR INTERACTIONS

As Figure 3.12 shows, the factors aforementioned can be categorised into four groups, namely in situ soil variation; material factors, which influence the miscibility and segregation tendency of the admixture; mixing equipment and workmanship. It

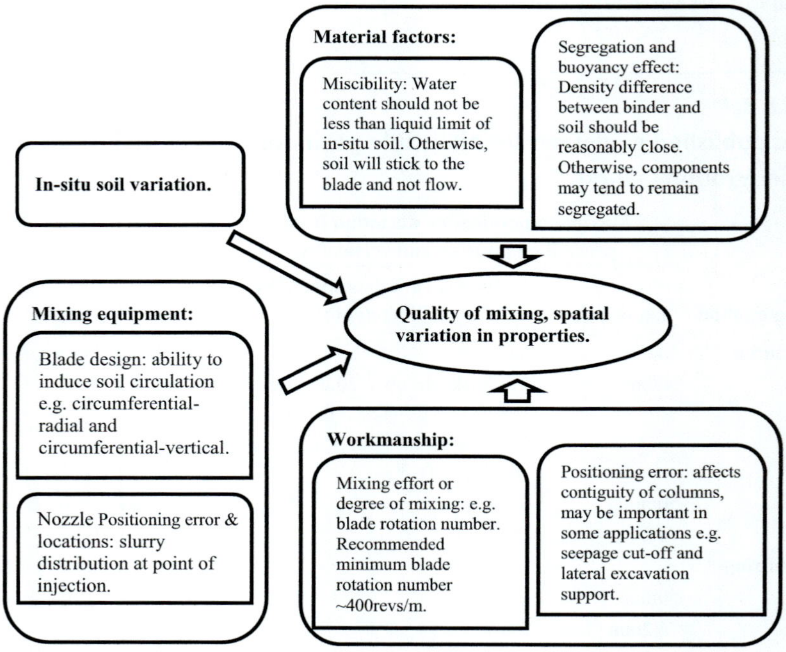

FIGURE 3.12 Summary of factors affecting the quality of mixing and heterogeneity of the improved soil mass.

is important to note that these factors are not completely independent. For instance, design of the mixing equipment, such as the locations of the injection nozzles, can affect the mixing effort required for adequate mixing. Similarly, the viscosity of the admixture and binder may influence the required mixing effort. Hence, a holistic view should be taken of the mixing equipment, materials to be mixed and the workmanship required.

REFERENCES

Akbas, S. O. and Kulhawy, F. H. (2010). Characterization and estimation of geotechnical variability in Ankara clay: A case history. *Geotechnical and Geological Engineering*, 28(5): 619–631.

Bahner, E. W. and Naguib, A. M. (1998). Design and construction of a deep soil mix retaining wall for the lake parkway freeway extension. *Soil Improvement for Big Digs: Proceedings of Sessions of Geo-Congress 98*, 18–21 October, ASCE, Boston, MA.

Bauer (2021). *CSM cutter soil mixing*. BAUER Maschinen GmbH, CSM 905.656.2 7.

Bellato, D., Marzano, I. P. and Simonini, P. (2017). A new parameter for the evaluation of mixing quality in deep soil mixing. *Proc. 5th Int. Conf. Grouting (Grouting 2017: Jet Grouting, Diaphragm Walls, and Deep Mixing)* (Eds. P. Gazzarrini, T. D. Richards, D. A. Bruce, M. J. Byle, C. S. El Mohtar and L. F. Johnsen), Honolulu, Hawaii: 375–384.

Bo, M. W., Arulrajah, A., Sukmak, P. and Horpibulsuk, S. (2015). Mineralogy and geotechnical properties of Singapore marine clay at Changi. *Soils and Foundations*, 55(3): 600–613.

Bouwmans, I., Bakker, A. and Van Den Akker, H. E. A. (1997). Blending liquids of differing viscosities and densities in stirred vessels. *Chemical Engineering Research and Design*, 75(8): 777–783.

Brocca, L., Morbidelli, R., Melone, F. and Moramarco, T. (2007). Soil moisture spatial variability in experimental areas of central Italy. *Journal of Hydrology*, 333: 356–373.

BS-EN (2001). *Execution of special geotechnical works—jet grouting*, British Standards Institution.

Burmester, S. S. H. (1993). *The mixing of miscible liquids with large differences in viscosity.* PhD thesis, University of Cambridge.

Chen, J., Liu, Y. and Lee, F. H. (2016). A statistical model for the unconfined compressive strength of deep-mixed columns. *Géotechnique*, 66(5): 351–365.

CDIT (2002). *The deep mixing method—principle, design and construction*, Coastal Development Institute of Technology, A. A. Balkema Publishers, Lisse/Abingdon/Exton (PA)/Tokyo, Japan.

Derksen, J. J. (2011). Blending of miscible liquids with different densities starting from a stratified state. *Computers and Fluids*, 50: 35–45.

Geo-Institute/ASCE (2016). *Jet grouting guide specification*, ASCE Geo-Institute Grouting Committee Jet Grouting Task Force (Rev. August).

He, L. (2018). *Spatial variability in deep soil mixing—centrifuge model and field data study.* PhD thesis, National University of Singapore: 220.

Herschel, W. H. and Bulkley, R. (1926). Konsistenzmessungen von Gummi-Benzollösungen. *Kolloid Zeitschrift*, 39(4): 291–300.

Honjo, Y. (1982). A probabilistic approach to evaluate shear strength of heterogeneous stabilized ground by deep mixing method. *Soils and Foundations*, 22(1): 23–38.

Horpibulsuk, S., Shibuya, S., Fuenkajorn, K. and Katkan, W. (2007). Assessment of engineering properties of Bangkok clay. *Canadian Geotechnical Journal*, 44: 173–187.

Huisman, J. A., Snepvangers, J. J. J. C., Bouten, W. and Heuvelink, G. B. M. (2002). Mapping spatial variation in surface soil water content: Comparison of ground-penetrating radar and time domain reflectometry. *Journal of Hydrology*, 269: 194–207.

Khatib, Z. and Richardson, J. F. (1984). Vertical co-current flow of air and shear-thinning suspensions of kaolin. *Chemical Engineering Research and Design*, 62: 139–154.

Kim, H. (2005). *Spatial variability in soils: Stiffness and strength*. PhD dissertation, Georgia Institute of Technology.

Larsson, S. (2005). State of practice report—execution, monitoring and quality control. *International Conference on Deep Mixing Best Practice and Recent Advances—Deep Mixing*, Stockholm.

Lee, F. H., Lee, C. H. and Dasari, G. R. (2006). Centrifuge modelling of wet deep mixing processes in soft clays. *Géotechnique*, 56(10): 677–691.

Lee, F. H., Lee, C. H., Dasari, G. R. and Zheng, J. (2008). Centrifuge study on uniformity of wet deep mixing. *International Journal of Physical Modelling in Geotechnics*, 8(1): 1–20.

Liu, Y., Lee, F. H., Quek, S. T., Chen, E. J. and Yi, J. T. (2015). Effect of spatial variation of strength and modulus on the lateral compression response of cement-admixed clay slab. *Géotechnique*, 65(10): 851–865.

Liu, W. F. and Leung, Y. F. (2018). Characterising three-dimensional anisotropic spatial correlation of soil properties through in-situ test results. *Géotechnique*, 68(9): 805–819.

Liu, Y., He, L. Q., Jiang, Y. J., Sun, M. M., Chen, E. J. and Lee, F. H. (2019). Effect of in-situ water content variation on the spatial variation of strength of deep cement-mixed clay. *Géotechnique*, 69(5): 391–405.

LTA (2020). *Engineering group materials & workmanship specification for civil & structural works, E/GD/09/104/A2, chapter 7—soil improvement works*, Land Transport Authority, Singapore.

Matsuo, T., Nisibayashi, K. and Hosoya, Y. (1996). Studies on soil improvement adjusted at low compressive strength in deep mixing method. *Proc. 2nd Int. Conf. on Ground Improvement Geosystems*, Tokyo, 1: 421–424.

Mizuno, Y., Namura, K. and Matsumoto, J. (1988). Field tests on deep mixing method as a solidification of sandy ground. *23rd Annual Meeting of the Japanese Soc. of Soil Mech. and Found. Engng.*: 2301–2304 (in Japanese).

Morey, J. and Campo, D. W. (1999). Quality control of jet grouting on the Cairo metro. *Ground Improvement*, 3: 67–75.

Ohtsubo, M., Egashira, K., Koumoto, T. and Bergado, D. T. (2000). Mineralogy and chemistry, and their correlation with the geotechnical properties of Bangkok Clay: Comparison with Ariake Clay. *Soils and Foundations*, 40(1): 11–20.

Passlick, T. and Doerendahl, K. (2006). Quality assurance in jet grouting for a deep seated slab in Amsterdam. *Piling and Deep Foundations*, Amsterdam.

Phoon, K. K. and Kulhawy, F. H. (1999). Characterization of geotechnical variability. *Canadian Geotechnical Journal*, 36(4): 612–624.

Puppala, A. J., Madhyannapu, R. J. and Nazarian, S. (2008). *Special specification for deep soil mixing*. The University of Texas, Arlington.

Raito. (2007). *RAS-JET system—mechanical soil mixing combined with jet grouting*. https://www.raito.co.jp/english/construction/pdf/ras-jet.pdf.

Reatto, A., Bruand, A., de Souza Martins, E., Muller, F., da Silva, E. M., de Carvalho Júnior, O. A. and Brossard, M. (2008). Variation of the kaolinite and gibbsite content at regional and local scale in Latosols of the Brazilian Central Plateau. *Comptes Rendus Geoscience*, 340: 741–748.

Rielly, C. D. and Pandit, A. B. (1988). The mixing of Newtonian liquids with large density and viscosity differences in mechanically agitated contactors. *Proc. 6th European Conf. on Mixing*, Pavia, Italy: 69–77.

Ryan, C. R. and Jasperse, B. H. (1989). Deep soil mixing at the Jackson lake dam. *Proc. ASCE Geotechnical and Construction Divisions Special Conference: Current Principles and Practices* (Ed. F. H Kulhawy). ASCE, New York, NY. June 1989: 354–367.

Rutherford, C. J., Biscontin, G. and Briaud, J. L. (2005). *Design manual for excavation support using deep mixing technology*. Texas A&M University.

Shibuya, S., Tamraka, S. B. and Theramas, N. (2001). Geotechnical site characterization on engineering properties of Bangkok clay. *Geotechnical Engineering Journal*, 32(3): 139–147.

Stoel, A. E. C. (2001). *Grouting for pile foundation improvement*. PhD thesis, Delft University of Technology.

Teo, H. J. (2017). *1-G Modelling of wet deep mixing with fibres*. PhD thesis, National University of Singapore.

Valdez-Llamas, Y. P., Auvinet, G. and Nunez, J. (2003). Spatial variability of the marine soil in the Gulf of Mexico. *Proc. OTC*, Houston, TX: OTC15266.

Xia, J., Cai, C., Wei, Y. and Wu, X. (2019). Granite residual soil properties in collapsing gullies of south China: Spatial variations and effects on collapsing gully erosion. *CATENA*, 174: 469–477. ISSN 0341-8162. https://doi.org/10.1016/j.catena.2018.11.015.

Yi, J. T., Li, Y. P., Bai, S., Fu, Y., Lee, F. H. and Zhang, X. Y. (2018). A simple effective stress approach for modelling rate-dependent strength of soft clay. *Journal of Offshore Mechanics and Arctic Engineering*, 140(4): 044501 (5 pp).

Yoshizawa, H., Okumura, R., Hosoya, Y., Sumi, M. and Yamada, T. (1997). JGS TC report: Factor affecting the quality of treated soil during execution of DMM. *Proc. 2nd Int. Conf. on Ground Improvement Geosystems*, Tokyo, 2: 931–937.

Zarlenga, A., Fiori, A. and Russo, D. (2018). Spatial variability of soil moisture and the scale issue: A geostatistical approach. *Water Resources Research*, 54: 1765–1780. https://doi.org/10.1002/2017WR021304.

4 Mechanical Properties and Behaviour of Stabilised Soils

4.1 CEMENT-TREATED SAND

Coarse sand cannot sustain pozzolanic reaction owing to the small surface-area-to-volume ratio arising from their large particle sizes. However, natural sandy soils often have small amounts of fines, which may contain pozzolanic materials such as silica. Asghari et al. (2003) reported pozzolanic hardening of admixtures of lime and a sandy alluvium containing about 6% fines. In addition, Nakarai and Yoshida (2015) reported that sand–lime admixtures exposed to the atmosphere may undergo long-term hardening well beyond 28 days, owing to carbonation of the lime by carbon dioxide in the air. However, this is inapplicable to deep mixing where the lime is not exposed to the atmosphere. The most commonly used hardening agent for sand is Ordinary Portland Cement (OPC). Unless some pozzolans are also added, the primary hardening reaction is hydration reaction. The microstructural changes of cement-treated sand are not dissimilar to those of concrete. Subramaniam et al. (2018) noted that 1-day cured samples of OPC-treated sand comprised loosely packed assemblages of calcium silicate hydrates with small amounts of needle-shaped ettringite crystal. The ettringite is likely to have been produced by the reaction between tricalcium silicate and the small amount of gypsum (calcium sulphate), which is added to prevent "flash setting." After 28 days, the hydration reaction is largely completed, and the calcium silicate hydrate assemblages are much more densely packed. By this time, the ettringite has been almost completely converted to monosulphate. Hence, the hardening effect of OPC on sand is derived mainly from the infilling of the large voids by cementitious products, mainly calcium silicate hydrate. This is similar to the gel-space theory (Powers 1958), which states that the strength of hardened cement paste is proportional to the power of the proportion of space infilled by calcium silicate hydrate.

Some blended cements such as Portland slag cement, such as CEM II/A–S and CEM II/B–S, as well as blast furnace cement, such as CEM II/A–C, which contains varying amount of ground granulated blast furnace slag (GGBS), are able to undergo pozzolanic reaction due to silica and alumina present in the GGBS. Nonetheless, the pattern of microstructural changes is likely to be similar, that is infilling of the voids by cementitious products.

Sand treated with calcium sulphoaluminate (CSA) cement produced significantly more ettringite over the first few days of curing, owing to the abundance of sulphate in the cement. This leads to higher initial strength compared to OPC-treated sand.

DOI: 10.1201/9781003285885-4

Subramaniam et al. (2018) also noted that, under wet-curing conditions, the strength of CSA-treated sand appears to reach a peak after a curing period of 7 days, followed by a slight decrease before regaining after about 14 days. They attributed this to the conversion of the ettringite to monosulphate, before the cementitious products are fully formed. However, upon completion of hydration reaction, the strengths of OPC and CSA cement-treated sand are largely equivalent. The final microstructures are also similar, comprising mainly cementitious products infilling the voids.

Chemical treatment of sand is usually conducted using binder content of about 10% or less and water content of about 20% to 30%, depending on the target strength of the improved soil. The relatively low water content reflects the much lower liquid limit of sand, compared to clay. As discussed in Chapter 3, for mixing to be feasible, the total water content of the admixture should be higher than the liquid limit of soil. Sand typically has much lower liquid limit than clay. Hence, mixing is often feasible at lower water content than with clay.

4.2 CEMENT-TREATED CLAY

Cement-treated clay differs from cement-treated sand in several aspects. As clay matrix has much smaller pores than sand matrix, the infilling effect is likely to be limited. Furthermore, during chemical stabilisation of clay, water is often added to raise the water content of the admixture to facilitate mixing. Hence, the water content at mixing is often higher than that of sand. Finally, clay particles often contain a significant amount of silica and alumina, which, if dissolved, can participate in pozzolanic reaction.

The initial reaction between cement and inorganic soft clay is still hydration reaction of cement. This produces rapid cementation, as well as lime, as a by-product. The latter raises the pH of the pore water, enabling the silicates and aluminates from the kaolinite go into solution. As discussed in Chapter 2, Fernandez et al. (2011) noted that silicate and aluminate from kaolinite sheets are more readily dissolved than those in illite since the former contains more hydroxyl groups.

Using scanning electron microscopy, Chew et al. (2004) noted that remoulded, untreated Singapore marine clay has an open type of microstructure, with stacks of platy clay particles in a dispersed arrangement. After treatment by 10% cement content and 28 days of curing, the open microstructure becomes more obvious due to the presence of large voids, with signs of reticulation. The increase in pore size can be attributed to the increase in water content required to facilitate mixing as well as a transformation from a dispersed to a flocculated microstructure. As the cement content increases to 50%, the pore size continues to increase. The increase in pore sizes is also reflected in pore size measurements using mercury intrusion porosimetry. Particle size deduced from pore size measurements as well as that measured using particle size analyser also shows an increase with cement content. However, for the same void ratio, the permeability of the cement-treated clay is lower than that of the untreated clay by roughly one order of magnitude. For the same sample, there is an initial increase in permeability, due probably to the addition of water which increases the void ratio. However, after 7 days of curing, the permeability decreases significantly.

Index tests also show a roughly commensurate rise in both liquid and plastic limits, with the latter rising slightly more than the former. This leads to a slight decrease in plasticity index. For lime-treated clay, Locat et al. (1996) noted similar trends of increase in the liquid and plastic limits. This indicates that a similar set of events is likely to occur with lime, albeit without the initial hydration-induced cementation process. Whereas the untreated marine clay would classify as a high to very high plasticity clay on the plasticity chart (e.g. ASTM 2011; EN ISO 14688–2 2018), the treated marine clay would classify as a very high plasticity silt. The transition from clay to silt reflects an increase in particle size during treatment, due to the agglomeration of small clay particles into large particle assemblages. However, as will be discussed, the classification of "high plasticity" for the treated clay may be deceptive.

Figure 4.1 shows clay–cement agglomerates with enclosed voids within the microstructure of intact cement-admixed marine clay. These agglomerates are also evident in the microstructure of remoulded cement-admixed marine clay (Figure 4.2), which indicates that the agglomerates are not completely pulverised during the remoulding process. Figure 4.3 shows the particle and pore sizes of remoulded cement-admixed marine clay. Cement treatment leads to an increase in particle size, which is consistent with agglomeration as well as the transition from clay- to silt-sized particles suggested by the Atterberg limits. On the other hand, pore opening sizes are roughly

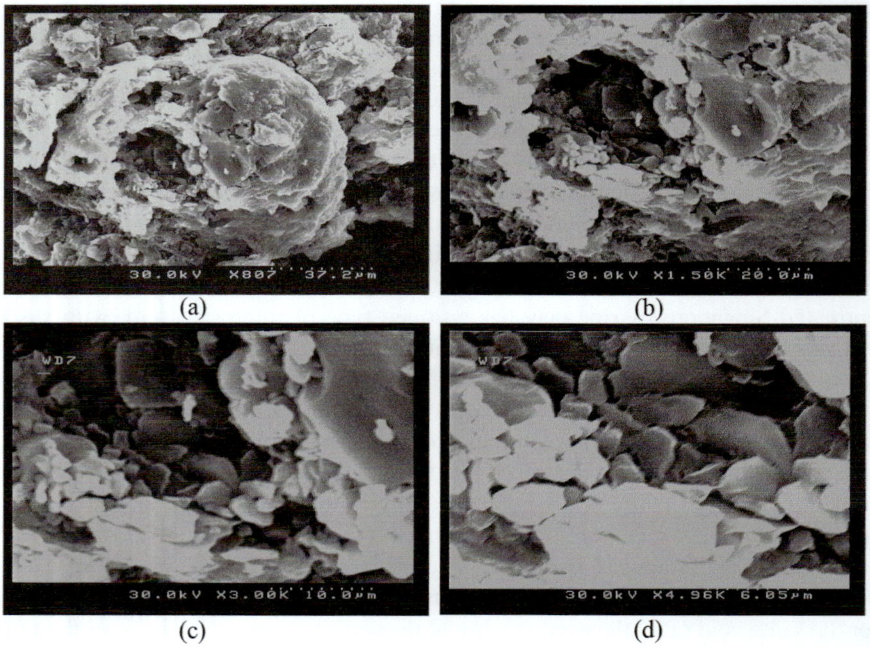

FIGURE 4.1 Post-hardened cement–clay agglomerates of much smaller clay particles with enclosed pore.

Source: **Chin 2006.**

FIGURE 4.2 Agglomerates in a remoulded cement-admixed marine clay specimen under (a) 3000 times and (b) 5000 times magnification.

Source: **Chin 2006.**

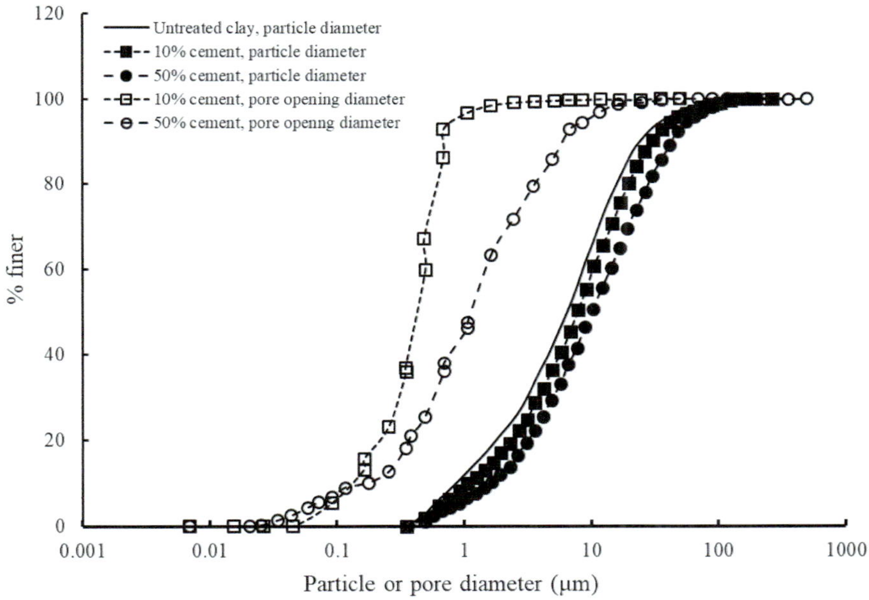

FIGURE 4.3 Particle and pore diameter of untreated and cement-treated marine clay after 7-day curing.

(*Data source:* Chew et al. (2004).

an order of magnitude smaller than the corresponding particle sizes. They are even smaller than the particle sizes of untreated marine clay. This suggests that cement treatment induces a reduction in pore opening sizes due to blockage by cementitious products, resulting in large enclosed pores with small openings, within the agglomerates. The microstructure is analogous to an egg shell with small openings. This allows water to remain entrapped within the enclosed pores, leading to a roughly commensurate increase in water content liquid and plastic limit measurements. The increase in liquid and plastic limits is due to the presence of unbroken agglomerates, even in the remoulded soil. The small pore openings also imply that the entrapped water is less mobile than that in the inter-agglomerate voids, leading to a decrease the permeability.

This agglomerate structure is induced by calcium ions adsorbed into the clay surfaces, giving rise to a flocculated clay structure. The mode of flocculation is likely to be edge-to-edge or edge-to-face rather than face-to-face, since the former would be more consistent with an open agglomerate microstructure with large enclosed pores. This flocculation process increases the effective size of the agglomerates while entrapping water within them. The dissolved silicate and aluminate ions react with the calcium ions to form cementitious products such as calcium silicate hydrate and calcium aluminate silicate hydrate. These cementitious products are deposited on or near the surfaces of the flocculated clay clusters, forming clay–cement agglomerates, thereby reducing the entrance pore diameter and lowering the permeability. Similar processes have also been reported for lime-stabilised marine clays (e.g. Rao and Rajasekaran 1996). Continued deposition of cementitious products on the surfaces of the clay agglomerates may also lead to encapsulation of the some of the remaining soil particles, thereby reducing their reactivity (e.g. Guo et al. 2022). Hence, the overall effect is one of progressive removal of kaolinite from the clay matrix and deposition of cementitious material onto the flocculated clay agglomerates. As a result of the hollow agglomerate microstructure, the void ratio and water content cement-admixed clay are usually significantly higher than those of cement-admixed sand. The agglomerates, which are silt-sized particles, are also usually weaker and more crushable than sand particles. All these have implications on the properties and constitutive behaviour of treated clay.

Chemical treatment of inorganic clay is usually conducted using binder content of up to about 20% and water content of about 50% to 100%, depending on the target strength of the improved soil and the plasticity of the soil. For highly plastic and organic soils, even higher binder content may be required.

4.3 STRENGTH

4.3.1 Cement-Treated Sand

The most commonly measured strength parameter in cement-treated soil is the unconfined compression strength, which is the maximum compressive strength of a cylindrical specimen of soil, typically with height-to-diameter ratio of 2. This is equivalent to the cylinder strength in concrete. Relationships for estimating the unconfined compressive strength of cement-treated sand or sandy soils are still relatively scarce. However, one may consider concrete to be an idealised form of cement-treated sand, although concrete often has a lower total water content (defined as ratio of weight of water to the total weight of soil and cement) than cement-treated sandy soil. A widely used relationship for estimating the strength of concrete is Abrams'

(1919) law, which states that the unconfined compressive strength q_u is related to the volumetric water–cement ratio of the mix $\frac{V_w}{V_c}$ by the following relationship:

$$q_u = \frac{A}{B^{\left(V_w / V_c\right)}} \tag{4.1}$$

Here A and B are fitted coefficients and V_w and V_c are the volume of water and dry cement. For q_u expressed in kPa, $A = 96527$kPa and $B = 7$. Abrams (1919) proposed that his relationship holds as long as the mix is workable, that is not too dry and the aggregate not too coarse. Abrams' law refers implicitly to the water–cement ratio at mixing, rather than the post-curing ratio. Since concrete mixes are often nearly saturated, V_w is also approximately the volume of voids V_v. In such cases,

$$\frac{V_v}{V_c} = \frac{\left(V_v / V_T\right)}{\left(V_c / V_T\right)} = \frac{\eta}{C_v} \tag{4.2}$$

where V_T is the total volume of the soil, η is the porosity and C_v is the volumetric cement content, that is the ratio of volume of cement to the total volume of the admixture. Hence, V_w / V_c can be expressed as η / C_v and Equation (4.1) can also be expressed as follows:

$$q_u = \frac{A}{B^{\left(\eta / C_v\right)}} \tag{4.3}$$

The phase relationship between soil, cement (or binder) and water at mixing is further explored in Figure 4.4, wherein s, c and w represent the masses of soil, cement and water, respectively; G_s (~2.7), G_c (~3.15) the specific gravities of soil and cement, respectively; and ρ_w the density of water. The volumetric water–cement ratio can now be expressed in terms of the mass water–cement ratio as

$$\frac{V_w}{V_c} = \frac{\left(w / \rho_w\right)}{\left(c / \left(G_c \rho_w\right)\right)} = G_c \frac{w}{c} \tag{4.4}$$

Hence, Equation (4.1) can be expressed as

$$q_u = \frac{A}{B'^{\left(w / c\right)}} \tag{4.5}$$

where $B' = B^{G_c} = 459.3$.

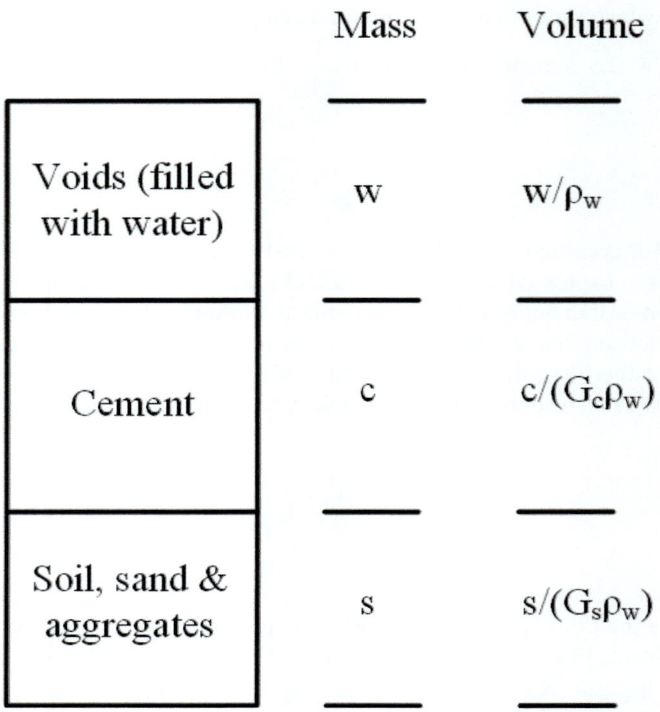

FIGURE 4.4 Phases in a saturated concrete or cement–soil mix.

Based on the phase quantities in Figure 4.4, the volumetric cement content C_v is given by

$$C_v = \frac{1}{G_c\left(w/c\right) + \left(G_c/G_s\right)\left(s/c\right) + 1} \tag{4.6}$$

In view of Equations (4.2) and (4.4) and assuming in a saturated mix, $V_v = V_w$.

$$G_c\left(w/c\right) = \eta\left\{G_c\left(w/c\right) + \left(G_c/G_s\right)\left(s/c\right) + 1\right\}$$

$$\Rightarrow \left(w/c\right) = \frac{e}{G_c}\left(\frac{A_w}{G_c/G_s + A_w}\right)^{-1} \tag{4.7a}$$

$$\text{and } \frac{V_w}{V_c} = e\left(\frac{G_c}{G_s A_w} + 1\right) \tag{4.7b}$$

Here e is the void ratio of the saturated mix and A_w is the cement content. Hence, both the volumetric and mass water–cement ratios, and thus Abrams' law, can be expressed in terms of porosity, or void ratio, and cement content.

Although Abrams' law has been proposed for application to cement-treated soils (e.g. Horpibulsuk et al. 2003), most adaptation of Abrams' law involves some modification. For cement-treated residual soil from weathered sandstone, Consoli et al. (2007) proposed that

$$q_u(\text{kPa}) = E\left\{\frac{\eta}{C_v{}^m}\right\}^{-n} \tag{4.8}$$

where $E = 5.03 \times 10^7$kPa, $m = 0.28$ and $n = 3.32$. Equation (4.8) differs from Equation (4.3) in two ways. Firstly, η and C_v appear in the base rather than the index. Secondly, the indices for η and C_v are different, indicating that the importance of the two parameters is different. Consoli et al. (2007) noted that using the same index for η and C_v gives a sub-optimal fit to the data and a better fit is obtained by according the porosity η a greater effect than C_v on the strength. For the saturated condition, Equation (4.8) can also be expressed in terms of w/c and s/c, as

$$q_u = E\left(\frac{\eta}{(C_v)^m}\right)^{-n} = E\left(\frac{\eta}{C_v}\right)^{-n}(C_v)^{-(1-m)n} = EG_c\left(w/c\right)^{-n}(C_v)^{-(1-m)n} \tag{4.9}$$

Combining Equations (4.6) and (4.9) leads to

$$q_u = E\left(w/c\right)^{-n}\left\{\left(\frac{w}{c}\right) + \frac{1}{G_s}\left(\frac{s}{c}\right) + \frac{1}{G_c}\right\}^{(1-m)n} \tag{4.10}$$

Equations (4.3) and (4.8) suggest that cement-treated sandy soil may differ from concrete, in terms of the relationship between strength and mix proportion. Construction-grade concrete typically has cement content of about 15% to 20%, water content of about 7% to 8% and porosity of 9% to 10%. On the other hand, Consoli et al.'s (2007) specimens have maximum cement content of 12%, water content ranging from 4% to 13.4% and porosity ranging from approximately 25% to 36%. Hence, Consoli et al.'s (2007) cement-treated sandy soil has a lower cement content and higher porosity than concrete.

4.3.2 Cement-Treated Clay

Compared to cement-treated sand, more relationships have been proposed for estimating the strength of cement-treated clay. Some of these relationships are summarised in Table 4.1. Most of the relationships use component ratios at mixing rather than cured conditions. This is probably to facilitate ease of application. Moreover, many can be considered to be modifications or generalisations of Abrams' law. For instance, Horpibulsuk et al.'s (2003) relationship [Equation (4.1–3a) in Table 4.1] is a direct extension of Abrams' law by incorporating a log-time function to account for strength development over time. On the other hand, many relationships are based on Kida et al.'s (1977) and Gallavresi's (1992) [Equation (4.1–1a) in Table 4.1] modification of Abrams law, which involves making the water–cement ratio $\left(w/c\right)$ the base, instead of index, in the exponential function. In Kida et al.'s (1977) form, the exponent for $\left(w/c\right)$ is taken to be 1. However, most subsequent relationships, including Gallavresi's (1992), used a fitted value for the exponent. Horpibulsuk et al. (2011) also used Gallavresi's (1992) form, with the incorporation of a log-time function for progressive strength development. The Japan Cement Association (2007) also recommended a similar version of this relationship, which includes a parameter F [see Equation (4.1–4) in Table 4.1] to account for the strength of the soil as $\left(w/c\right) \rightarrow \infty$. However, F does not represent the strength of the untreated soil, since the soil is remoulded during admixing. Instead, it represents the strength of the treated soil as $\left(w/c\right) \rightarrow \infty$, which is likely to be negligible.

Many subsequent studies have shown that, apart from the water–cement ratio, the ratio of the mass of binder (c) to the mass of dry soil (s), termed as the binder content $A_w = c/s$, or its reciprocal, the soil–cement ratio s/c, also affects the strength of the treated soil. An example of this is Lee et al.'s (2005) relationship, which was subsequently extended by Xiao et al. (2014) to include the effect of time on strength development; see Equations (4.1–5a) and (4.1–5b) in Table 4.1. Similarly, Chian et al.'s (2016) relationship [Equations (4.1–7a) and (4.1–7b) in Table 4.1] incorporates s/c, using a linear, instead of exponential, relationship. Furthermore, as previously discussed, Consoli et al.'s (2007, 2016) relationship [Equation (4.8)] can also be expressed in terms of w/c and s/c.

Tang et al.'s (2001), Tsuchida and Tang's (2015) and Yamashita et al.'s (2020) relationships [Equations (4.1–6a) and (4.1–6b) and (4.1–8) in Table 4.1] also incorporate the effect of cement as well as water contents, or porosity. However, the departure from Abrams' and Gallavresi's forms is greater since the functional forms are not multiplicative.

This discussion indicates that, notwithstanding the diversity of empirical relationships, there are two important factors affecting the strength of porous cement-admixed clay, namely the volumetric or mass water-binder ratio, that is V_w/V_c or $\left(w/c\right)$, and the binder content or soil–binder ratio, that is A_w or s/c. However, since the water–cement (or binder) ratio is also a function of the void ratio and binder content for saturated mix, it is possible to re-classify the influencing factors into a geometric factor, represented by the void ratio, and a material factor, represented by the binder content. This notion is similar to that proposed by Lim et al. (2021) for chemically treated sludge with and without sand. Using data from ten studies involving sludge with widely different sand content stabilised by a various binders including OPC, lime, sulphoaluminate cement and other additives, Lim et al. (2021) proposed that the upper strength limit of the treated sludge varies inversely with the parameter $\left(e/A_w^*\right)$ in which A_w^*, the binder–sludge ratio, is akin to the binder content,

TABLE 4.1
Some proposed relationships for estimating strength of cement-treated clay.

Reference	Relationship	Remarks
Gallavresi (1992)	$q_u = \frac{q_0}{\left(w/c\right)^n}$ (4.1–1a) Fitted parameters: $n = 1.25–3$, q_0 varies depending on cement and soil types and curing period. In terms of void ratio e and binder content A_w, $q_u = q_0 G_c^n e^{-n} \left(\frac{A_w}{G_c/G_s + A_w} \right)^n$ (4.1–1b) where G_c and G_s are the specific gravities of cement and soil, respectively.	This is a modified version of Abrams' law. However, the notion that strength depends only on the water–cement ratio remains the same.
Consoli et al. (2007, 2016)	$q_u = A \left\{ \frac{\eta}{C_v{}^m} \right\}^{-n}$ (4.4) where η is the porosity and C_v is the volumetric cement content. Fitted parameters: $A = 1.29 \times 10^9$ to 9.25×10^9kPa, depending upon the binders, soil type and curing period. $m = 0.12$ and 0.28 for lime- and cemented-treated dispersive clay, respectively, $n = 2.22$ for silty sand and 4.13 for dispersive clay. In terms of void ratio and binder content, $q_u(kPa) = E\left\{ e(1+e)^{m-1} \right\}^{-n} \left\{ \frac{A_w}{\left(\frac{G_c}{G_s} + A_w \right)} \right\}^{mn}$ (4.1–2) where E is fitted parameter.	As discussed in section 4.3.1, this is actually a modification of Abram's law as well as Gallavresi's (1992) relationship.

(*Continued*)

TABLE 4.1 ***(Continued)***

Some proposed relationships for estimating strength of cement-treated clay.

Reference	Relationship	Remarks
Horpibulsuk et al. (2003, 2011)	$$\frac{q_u\left(\frac{w}{c}\right)_{1,t}}{q_u\left(\frac{w}{c}\right)_{28}} = B^{\left[\left(\frac{w}{c}\right)_{28} - \left(\frac{w}{c}\right)_{t}\right]}(E + F\ln t) \quad (4.1\text{–}3a)$$ Fitted parameters: $B = 1.24$, $E = 0.038$, $F = 0.281$. $$\frac{q_u\left(\frac{w}{c}\right)_{t}}{q_u\left(\frac{w}{c}\right)_{28}} = \left\{\frac{\left(\frac{w}{c}\right)_{28}}{\left(\frac{w}{c}\right)_{D}}\right\}^{n}(E + F\ln t) \quad (4.1\text{–}3b)$$ Fitted parameters: $n = 1.27$, $E = 0.039$, $F = 0.281$.	Equation (4.1–3a) is a generalisation of Abrams' law proposed to fit strength development in Ariake and Bangkok clays for different curing periods t (in days). Equation (4.1–3b) is similar to Equation (4.1–3a), but is based on generalisation of Gallavresi's (1992) version of Abrams' law, instead.
Kida et al. (1977) and Japan Cement Association (2007)	$$q_u = \frac{q_0}{\left(\frac{w}{c}\right)^{n}} + F \quad (4.1\text{–}4)$$ Fitted parameters are q_0, n and F. In Kida et al. (1977), $n = 1$	This is a generalisation of Gallavresi's (1992) relationship, with a constant F included to represent strength of the soil as $\left(\frac{w}{c}\right) \rightarrow \infty$.
Lee et al. (2005) and Xiao et al. (2014)	$$q_u = \frac{q_\infty \exp\left(\frac{m}{A_w}\right)}{\left(\frac{w}{c}\right)^{n}}\left\{1 - \frac{1}{1 + \left(\frac{\alpha t}{q_\infty}\right)^{r}}\right\} \quad (4.1\text{–}5a)$$ $A_w = \left(\frac{c}{s}\right)$ is the cement or binder content. Fitted parameters: $n \sim 3$, $q_\infty \sim 40{,}000$kPa (q_∞ is obtained by setting t and A_w to ∞ and w/c to 1. Hence, it is equivalent to the long-term strength of cement paste with water–cement mass ratio of 1). $m = 0.2$–0.38 for OPC and untreated soil with plasticity index (PI) = 30–55%, = 0.48–0.6 for OPC and untreated soil PI = 55%–60%, for Upper and Lower Singapore marine clay.	This is an extension of Gallavresi's (1992) which includes the effect of soil–cement content $A_w = \left(\frac{c}{s}\right)$ and curing time t in days.

	α ~ 1300–2200kPa/day and r ~ 0.52–0.67 for Upper Singapore marine clay. These two parameters control the rate of hardening. Initial rate of increase in strength = αr. In terms of void ratio and binder content, $$q_u = \frac{q_\infty G_c{}^n}{e^n} \exp\left(\frac{m}{A_w}\right) \frac{A_w^n}{\left(\frac{G_c}{G_s} + A_w\right)^n} \left\{ 1 - \frac{1}{1 + \left(\frac{\alpha t}{q_\infty}\right)^r} \right\} \quad (4.1\text{–}5b)$$
Tsuchida and Tang (2015), Yamashita et al. (2020)	$$q_u = \left(E + F\ln(t)\right)\left(C^* - C_0^*\right)(1-\eta)^n \quad (4.1\text{–}6)$$ $C^* = 100\frac{c}{s+c}\% = 100\frac{A_w}{1+A_w}\%$ is the ratio of mass of cement to total mass of soil solid (in %), termed "added cement rate," and η is the porosity at mixing. Fitted parameters: C^*_0 is the threshold "added cement rate" below which there is no evident increase in strength (see threshold binder content in sub-section 2.5.3), N ~ 2.3–4.2, E ~ −1350–364 and F ~ 1890–3080 for different clays treated by cement. In terms of void ratio and binder content, $$q_u = 100\left(E + F\ln(t)\right)\left(\frac{A_w}{1+A_w} - \frac{A_{w0}}{1+A_{w0}}\right)(1+e)^{-n} \quad (4.1\text{–}6b)$$

(*Continued*)

TABLE 4.1 *(Continued)*

Some proposed relationships for estimating strength of cement-treated clay.

Reference	Relationship		Remarks
Chian et al. (2016)	$$q_u = \frac{E + F\left(s/c\right)}{B'^{\left(w/c\right)}} \ln(t)$$	(4.1–7a)	This is a generalisation of Abrams' law to include the effects of soil–cement mass ratio $\left(\frac{s}{c}\right)$ and curing time t.
	Fitting parameters: $E \sim 17.83\text{LL} - 346.46$ where LL is the liquid limit $F \sim 2.332 - 7.586 \times 10^{-3}E$ $B' = 1.89 - 0.00629\text{LL}$ In terms of void ratio and binder content,		
	$$q_u = \frac{E + F\left(1/A_w\right)}{B'^{\left[\frac{e}{G_c}\left(\frac{G_c}{G_s A_w}+1\right)\right]}} \ln(t)$$	(4.1–7b)	
Tang et al. (2001)	$$q_u = \frac{K\left\{\left(c/s\right) - \left(c/s\right)_0\right\}}{\left\{G_s\left(w/s\right)+1\right\}} = \frac{K\left\{A_w - A_{w0}\right\}}{\left\{G_s A_w\left(w/c\right)+1\right\}}$$	(4.1–8)	This relationship was developed for dredged clay with high water content.
	G_s is the specific gravity of the soil solid. Fitting parameters: $\left(c/s\right)_0$ is a threshold cement content, such that $\frac{\left(c/s\right)_0}{\left\{G_s\left(w/s\right)+1\right\}} = 33\text{kg/m}^3 - 62\text{kg/m}^3$ for different dredged clays. $K \sim 98\text{kPa/kg/m}^3 - 745\text{kPa/kg/m}^3$ for different clays and curing periods.		

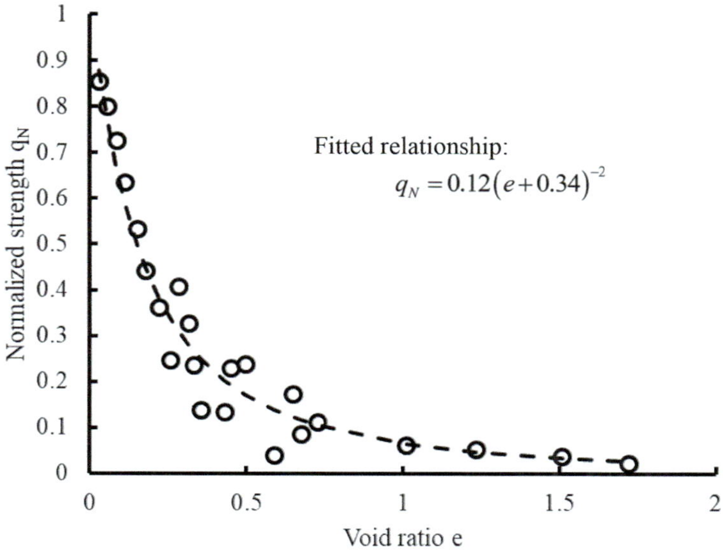

FIGURE 4.5 Decrease in strength of zirconia ceramic with void ratio. The compressive strength is normalised by the corresponding value at void ratio of 0. [Data source: Kulkov et al. 2003]

excluding the sand. The effect of porosity on the strength of cemented porous media has also been studied. Figure 4.5 shows Kulkov et al.'s (2003) data on porous zirconia ceramic under uniaxial compression replotted in terms of void ratio. As can be seen, the rate of decrease in strength is well fitted by an inverse power law similar to that for the water–cement ratio.

The strength relationships can be readily expressed in terms of void ratio e and cement content A_w. For instance, combining Equations (4.1) and (4.7b) allows Abram's law to be expressed as

$$q_u = \frac{A}{B^{\left\{e\left(\frac{G_c}{G_s A_w}+1\right)\right\}}} \tag{4.11}$$

Similarly, Gallavresi's (1992) relationship can be expressed as follows:

$$q_u = \frac{q_0 G_c^n}{\left\{e\left(\frac{G_c}{G_s}+A_w\right)\right\}^n} \tag{4.12}$$

4.3.3 Calibrating and Fitting Strength Relationships

Published fitted relationships, such as those in Table 4.1, are unlikely to be universally applicable. For this reason, a practical course of action may be to re-calibrate an existing relationship or synthesise a new relationship if a new type of soil(s) is to be treated. In such cases, judicious choice of the type of relationship to be calibrated or synthesised may yield significant savings in time and effort. The previous discussion shows that there are generally two major factors which need to be accounted in the fitting attempts, namely void ratio or porosity (a geometric factor representing the effect of voids) and binder content (a material factor). Virtually all the relationships in Table 4.1 can be expressed in terms of e and A_w, and most can be expressed in the form

$$q_u = \Psi F(e) G(A_w) \tag{4.12}$$

where Ψ is a constant, $F(e)$ and $G(A_w)$ are functions of e and G, respectively. $F(e)$ typically decreases with e and is quite often represented by a negative power function, whereas $G(A_w)$ typically increases with A_w. This reflects the opposing effects of void ratio and binder content. If a negative power function of void ratio e is used, $q_u \to \infty$ as $e \to 0$. To avoid this, a negative power function of $(e+C)$, in which C is a positive constant, similar to the fitted relationship in Figure 4.5, may be preferred.

Calcium silicate hydrate (C-S-H) is a hierarchical material (Constantinides and Ulm 2004). At a scale ranging from 1nm to 100nm, termed Level 0, C-S-H is presented as globules with non-regular molecular structures. Pellenq et al. (2009), using molecular dynamics simulation, predicted tensile strength of Level 0 (1nm to 100nm) CSH globules to range from 1GPa to 3GPa. The compressive strength is likely to be even higher. At Level 1 (100nm–10μm), these globules form assemblages with intervening pore space. Shahrin and Bobko (2017) reported compressive strength of Level 1 (100nm to 10μm) C-S-H micropillars ranging from 225MPa to 606MPa. At Level 2 (10μm to 1mm), the C–S–H assemblages form a hardened cement paste, which include unreacted lime, unhydrated cement clinkers, water and micron-sized pores. Hence, as the void ratio tends to zero, the material strength of C-S-H is far higher than the typical strength of chemically stabilised soil, or even concrete. As such, the errors of using a negative power law of void ratio to fit much lower strength levels may be insignificant.

The procedure for calibration varies between relationships and some are more readily calibrated than others. For instance, Xiao et al.'s (2014) relationship can be calibrated in two steps. The first step involves determining the unconfined compressive strength q_u from several mix ratios (typically ~8) after a constant curing duration, say 28 days. Since t = 28days, Equation (4.1–5) can be written as

$$q_u = \frac{q_\infty \exp\left(\frac{m}{A_w}\right)}{\left(\frac{w}{c}\right)^n}\left\{1 - \frac{1}{1+\left(\frac{\alpha t_{28}}{q_\infty}\right)^r}\right\} = \frac{q_{28} \exp\left(\frac{m}{A_w}\right)}{\left(\frac{w}{c}\right)^n} \tag{4.13}$$

where t_{28} = 28days and q_{28} is the 28-day strength for w/c = 1 and binder content of infinity. Equation (4.13) can be written as

$$\ln q_u + n\ln\left(\frac{w}{c}\right) = \ln q_{28} + m\left(\frac{1}{A_w}\right) = \ln q_{28} + m\left(\frac{s}{c}\right) \tag{4.14}$$

The value of n is then adjusted iteratively until the plot of $\ln q_u + n\ln\left(\frac{w}{c}\right)$ against $\left(\frac{s}{c}\right)$ gives a good linear fit. The values of m and $\ln(q_{28})$ are then given by the gradient and y-intercept of the fitted line. Experience with the Singapore marine clay indicates that $n \sim 3$ usually gives a good fit.

The rate of hardening parameters α and r, as well as q_∞, can be determined by measuring the strength q_u of several specimens with the same mix ratio, say R, over different curing durations t. Equation (4.13) can be rewritten as

$$q_u = q_R\left\{1 - \frac{1}{1+\left(\frac{\alpha t}{q_\infty}\right)^r}\right\} \tag{4.15}$$

$$\text{where } q_R = \frac{q_\infty \exp\left(\frac{m}{A_w}\right)}{\left(w/c\right)^n} \tag{4.16}$$

Equation (4.15) can be expressed as follows:

$$\ln\left(\frac{q_u}{q_R - q_u}\right) = r\ln\alpha + r\ln\frac{t}{q_R} \tag{4.17}$$

Plotting $\ln\left(\frac{q_u}{q_R - q_u}\right)$ versus $\ln\frac{t}{q_R}$ and adjusting q_R until a good linear fit is obtained allows q_R to be fitted and thus q_∞ to be evaluated. The gradient of the line is equal to r and the intercept is $r\ln\alpha$. Since q_R affects both the x- and the y-variables, the Deming regression, which measures the goodness of fit in terms of the square of the normal distances between the points and the fitted line, may be more appropriate than the standard linear regression, which uses the y-offset only.

It should also be noted that fitted strength relationships may only be valid within the fitted data range. For instance, as discussed earlier, relationships which use negative exponential function of void ratio will predict infinite strength as void ratio tends to 0. Similarly, relationships which use the log time function to describe strength development may yield unreasonable results, since the strength tends to $-\infty$ and ∞, as time tends to 0 and ∞. Chen et al. (2016) also noted that Lee et al.'s (2005) relationship, that is Equation (4.1–5a), may yield anomalously high strength at low cement content, owing to the exponential function. They suggested replacing the exponential function by a quadratic function, leading to the following equation:

$$q_u = \frac{q_\infty \left\{1 + \frac{m}{A_w} + \left(\frac{m}{A_w}\right)^2\right\}}{\left(w/c\right)^n} \left\{1 - \frac{1}{1 + \left(\frac{\alpha t}{q_\infty}\right)^r}\right\} \tag{4.18}$$

4.3.4 Long-Term Strength

The long-term strength of cement-treated soil has been investigated, both using laboratory samples (e.g. Terashi et al. 1983; Kitazume et al. 2003; Nakamura and Kitazume 2006) as well as core samples from in situ treated soil (e.g. Hayashi et al. 2003; Ichiba et al. 2002; Ohishi et al. 2002; Ikegami et al. 2006). In these studies, specimens which have been aged up to 26 years were tested. The general finding is that, away from the boundaries of the treated domain, there was no deterioration in strength. Indeed, strength increase was observed up to 10 to 20 years post-installation. On the other hand, soil close to the treated boundary decreased in strength due to de-calcification by leaching. However, Ikegami et al. (2006) reported that strength decrease is restricted to a thin zone about 30mm adjacent to the domain boundary. Elsewhere, there was no deterioration in strength. A similar trend was also reported for lime-treated clay (Kitazume and Takahashi 2009).

4.3.5 Tensile Strength

Tensile strength is commonly measured using the split tensile test (also known Brazilian tensile test). Some measurements have also been made using bending and direct tensile tests. However, Terashi et al.'s (1980) data showed that tensile strength measurements using bending tests are significantly more scattered than the split tensile tests. Tensile strength ratio in bending tests have also been observed to reach as high as 0.6 times the unconfined compressive strength. Namikawa and Koseki (2007) attributed this over-estimation to strain softening on the tensile side of the beam. Namikawa and Koseki (2007) also compared measurements from direct tensile and split tensile tests. These measurements indicate that direct tensile and split tensile tests give similar scatter, but the former generally gives higher tensile-to-unconfined-strength ratio than the latter. Namikawa and Koseki (2007) attributed this to shear yielding at the loading point of the split tensile tests. However, direct tensile tests are generally considered to be more challenging to conduct as they required appropriate end grips on the specimens.

As Table 4.2 shows, over a wide range of materials, ranging from sand to clay, the ratio of tensile to unconfined compressive strength is bounded between 0.12 and 0.35 with the exception of Liu et al. (2022), who reported a relatively low ratio of 0.07 for cement-treated Champlain sea clay, a sensitive clay.

TABLE 4.2
Some reported tensile-to-compressive strength ratios.

Reference	Material	Split tensile:Unconfined compressive strength ratio
Terashi et al. (1980)	Lime and cement-treated clay	~0.15 for unconfined compressive strength up to 2MPa, decreasing thereafter.
Saitoh et al. (1996)	Cement-treated sand and clay	0.05–0.35
Koseki et al. (2005)	Cement-treated sand	0.15–0.24
Consoli et al. (2010)	Cement-treated Osorio sand (a uniform fine sand)	0.15
Tariq and Maki (2014)	Cement-treated uniform sand	0.12–0.16
Pan et al. (2016)	Cement-treated highly plastic marine clay	0.12–0.16
Liu et al. (2022)	Cement-treated Champlain Sea clay	0.07

4.3.6 Failure Envelope

Table 4.3 summarises the results from some studies on the failure envelopes of cement-treated soil. Compared to the unconfined compressive strength, published literature on the failure envelopes are scarcer and the majority of the studies deal with cement-treated sand. The angle of friction and effective cohesion are still not well correlated to the mix proportion or void ratio. For cement-treated sand, reported angles of friction range from 32° to 56°, but no clear trend of change with binder content or void ratio has been reported. Effective cohesion typically increases with binder content.

For cement-treated clay, studies on the Mohr-Coulomb failure envelope parameters appear to be even fewer, the unconfined compressive strength being more often used in design. Reported angles of friction range from 28° to 35° for cement-treated high plasticity soft clay (Uddin et al. 1997; Chin 2006). This range is significantly higher than the angle of friction of the untreated clay and indicates that cement treatment is likely to increase the angle of friction of soft clay. The effective cohesion also increases with cement content.

One of the difficulties in interpreting Mohr Coulomb peak strength parameters of chemically stabilised soil lies in their strain softening behaviour, both under drained and undrained conditions. Figure 4.6 shows the stress–strain curves of undrained and drained tests on Singapore marine clay treated with cement. In the specific volume versus mean effective stress plots of this figure, the specific volume is equal to (1+void ratio) and is defined as the total volume of the soil divided by the volume of soil solids.

The peak strength in the drained tests is consistently reached at a significantly higher strain level than that in the corresponding undrained tests. The reason for this is that the undrained stress paths trends towards the tension cut-off line, which is a

TABLE 4.3
Failure envelope characteristics from previous studies.

Reference	Material	Failure envelope: Cohesion, c' (kPa)	Failure envelope: Angle of friction, ϕ' (°)	Remarks
Uddin et al. (1997)	Soft clay	$c' \sim 15$kPa (5% cement) and 215kPa (10% cement)	$\phi' \sim 31°$ (5% cement) and 28° (10% cement)	High plasticity clay treated with 5% and 10% cement. Strength parameters inferred from maximum deviator stress from undrained stress paths.
Chin (2006)	Soft marine clay	$c' \sim 160$kPa[1]	$\phi' \sim 35°$[1]	High plasticity clay treated with 20% cement. Curing period 7 days. Strength parameters inferred from maximum deviator stress from undrained stress paths.
Muhunthan and Sariosseiri (2008)	Everett glacial till	100	35°	Coarse-grained soil treated with 5% cement
	Aberdeen Marine deposit	60	29°	Fine-grained medium plasticity soil treated with 5% cement
Consoli et al. (2012)	Cement-treated uniform fine sand	A parabolic failure envelope is proposed based on the parameter $\left(\frac{\eta}{C_v}\right)$, instead of c' and ϕ'.		Cement content 3% to 10%. Void ratio of treated sand 0.71 to 0.82.
Ajorloo et al. (2012)	Cement-treated loose medium sand with sub-angular grain	Peak strength $c_p' = 66$ to 1189kPa, increasing with cement content and curing duration. Critical state $c_p' = 4$–80kPa.	Peak strength $\phi_p' = 32°$ to 50°, increasing with cement content and curing duration. Critical state $\phi_p' = 34°$–56°.	Binder is a mixture of cement, lime and silica fume in different proportions. Binder content 4% to 32%. Curing duration 28 and 180 days.
Consoli (2014)	Uniform fine sand and silty sand	$c' = 57$kPa–346kPa Proposed relationship $c'(kPa) = 7365\left(\frac{\eta}{C_v}\right)^{-1.3}$	$\phi' = 27°$ to 44°	Cement content 1% to 5%. Curing period 7 days.
Rios et al. (2014)	Silty sand	Ultimate $c' = 0$kPa	Ultimate $\phi' = 46°$ (for effective confining stress < 1000kPa) and 29° (for effective confining stress from 4000kPa to 14,000kPa)	Cement content 2% to 7%. Curing period 7 days.
Shooshpasha and Shirvani (2015)	Cement-treated sandy soil	Not available	Peak strength angle of friction $\phi_p' =$ 36° to 50°, increasing with cement content and curing duration	Cement content 2.5% to 7.5%. Curing duration 7 to 28 days.

[1] Actual peak strength envelope is slightly curvilinear. These strength parameters were estimated from a linearised failure envelope.

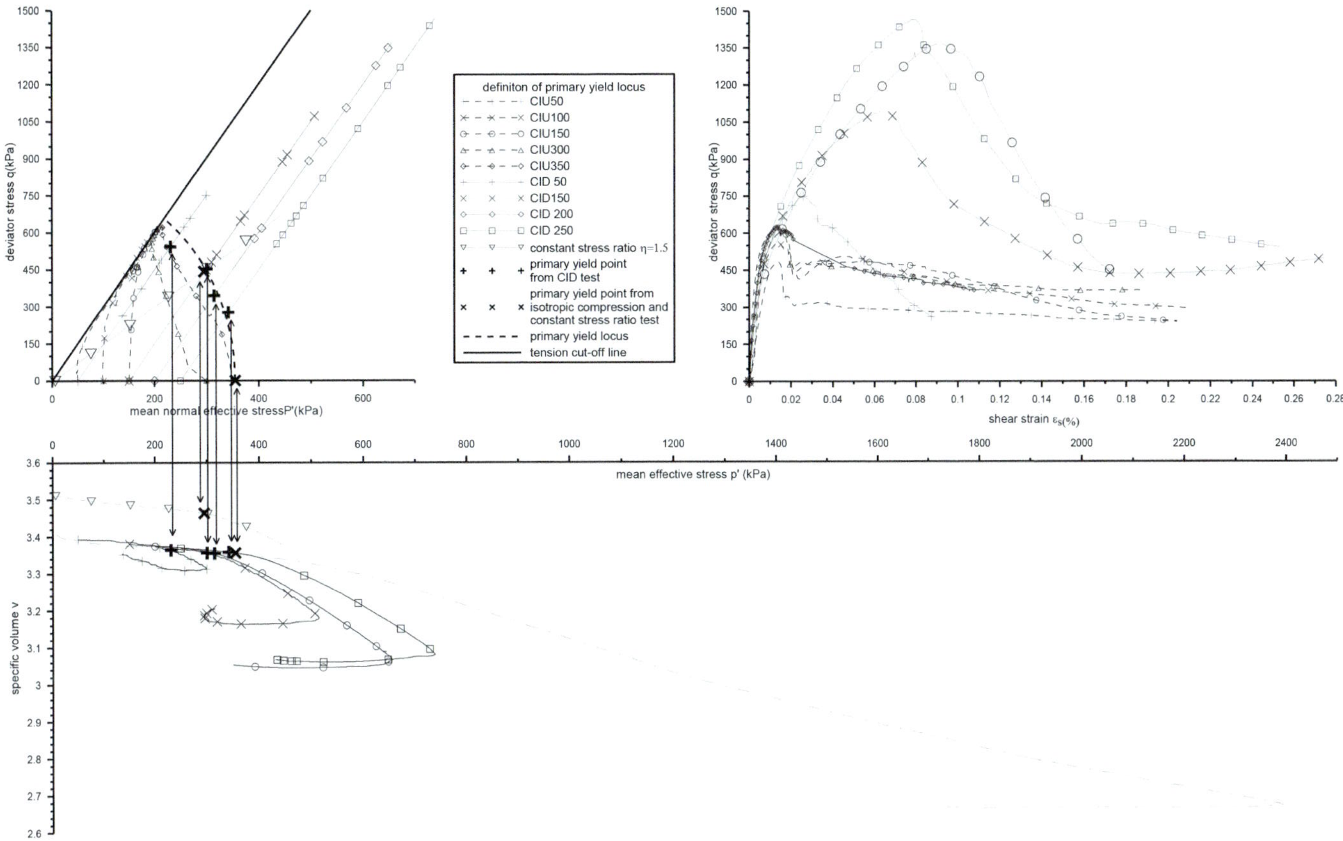

FIGURE 4.6 Stress paths, compression and stress–strain curves as well as fitted yield locus for Singapore marine clay treated with cement content 30% and water content 100%, cured under atmospheric pressure for 7 days. In the test identifier, CIU and CID represent consolidated undrained and drained tests, respectively, and numeral suffixes denote effective confining stress.

Source: **Xiao et al. 2014.**

line passing through the origin in p′-q stress space, with a gradient of 3. All points on this line correspond to states with zero effective lateral stress. The region to the left of the tension line represents stress states with negative effective lateral stress. However, this region is unattainable in triaxial test situations since the pore pressure within the sample will exceed the total lateral stress, and pore water will flow into the interface between sample and rubber membrane, thereby violating the undrained condition. When this occurs, the membrane no longer provides confinement and the sample is akin to an unconfined compression condition. The peak strength of the undrained samples is typically reached in the vicinity of the tension cut-off line, before the fitted yield locus is reached. Loss of cementation generally occurs after onset of yielding. Hence, the undrained peak strength is likely to comprise a large cohesion component arising from cementation.

On the other hand, the peak strength of drained samples is reached well after the yield locus is exceeded. Thus, significant loss of cementation is likely to have occurred when the sample reaches peak strength under drained condition. Xiao et al. (2017) postulated that the peak strength in drained tests is due to a decrease in the angle of friction as cemented soil agglomerates break down under plastic strains, resulting in slip plane formation and localisation behaviour. As Figure 4.6 shows, the compression curves of drained samples show decrease in mean effective stress post-peak, but the specific volume remains largely unchanged, indicating an absence of dilation. By this stage, a lot of the initial cementation would have been destroyed. Hence, peak strength envelope in drained tests is likely to be associated with a smaller cohesion than that in undrained tests. Amini and Hamidi's (2014) data on the discrepancies between the peak strength envelopes from drained and undrained tests indicate similar phenomenon in cement-treated sand. However, the drained peak strength in cement-treated sand is likely to be associated more with dilatancy behaviour, causing an increase in specific volume or void ratio, rather than with particle breakdown. Hence, care must be exercised in correlating the peak strength from undrained and drained tests on chemically stabilised soils.

4.4 MODULUS

4.4.1 Secant and Maximum Modulus

Much of the early studies on modulus of cement-treated soil were conducted on secant modulus. Table 4.4 shows the secant Young's modulus measured at about 50% of peak strength, E_{50}, from some studies. The secant modulus is usually measured from unconfined compression tests and is often expressed as a multiple of the unconfined compressive strength, termed herein as rigidity index $\left(E_{50}/q_u\right)$. Many of the studies appeared to focus on fine-grained soils such as silt and clay. Studies on sand appear to be scarcer. The rigidity index varies over a wide range, from as low as 50 to as high at 1000. This variability may be explained by the differences in the untreated soil properties, test set-ups and methodology, as well as stress levels, as a percentage of the strength, at which the modulus is measured. The effect of confining stress on E_{50} has not been well studied.

TABLE 4.4
Secant Young's modulus measured at 50% strength (E_{50}) from some studies.

Reference	In situ soil type	Binder type	E_{50}-to-q_u relationship
Terashi et al. (1977)	Kawasaki Clay and a marine clay	Lime	Kawasaki clay: $E_{50} \sim 200q_u$ to $1000q_u$ Honmoku marine clay: $E_{50} \sim 75q_u$ to $200q_u$
Saitoh et al. (1980)	Silty clay	Cement	$E_{50} \sim 350q_u$ to $1000q_u$
Kawasaki et al. (1981)	Clays and sandy silt (16 soil types)	Cement	$E_{50} \sim 350q_u$ to $1000q_u$
Geotesting Express (1996)	Boston Blue Clay	Cement	$E_{50} \sim 50q_u$ to $150q_u$
Asano et al. (1996)	High plasticity silty clay	Cement and a blend of coal fly ash, gypsum and cement	$E_{50} \sim 140q_u$ to $500q_u$ (cement-treated) $E_{50} \sim 50q_u$ to $300q_u$ (treated using a blend of coal fly ash, gypsum and cement)
Futaki et al. (1996)	Organic soil	Cement	$E_{50} \sim 140q_u$ to $173q_u$
Lee et al. (2005)	High plasticity marine clay	Cement	$E_{50} \sim 80q_u$ to $200q_u$

As Table 4.5 shows, much of the more recent works deal with the small strain modulus, which is also maximum modulus. The maximum shear modulus G_{max} is more commonly measured than the Young's modulus, although they are interconvertible if the Poisson's ratio is known. In many studies (e.g. Tanaka and Terashi 1986; Kamruzzaman 2002; Puppala et al. 2006; Tatsuoka 2010; Verástegui Flores et al. 2010; Seng and Tanaka 2011; Kang et al. 2017), G_{max} is also expressed as a multiple of the strength. In these studies, G_{max} is typically measured using bender elements in the absence of confining stresses. The influence of effective stress on G_{max} has not been widely studied.

Hoyos et al. (2004) and Trhlikova et al. (2012) used modifications of Viggiani and Atkinson's (1995) relationship to express the effect of effective stress on G_{max}; see Equations (4.5–2) and (4.5–7) in Table 4.5. Trhlikova et al. (2012) also included the effect of structure in a multiplicative format. However, Hoyos et al.'s (2004) relationship will predict $G_{max} = 0$ as $p' = 0$, which is counter-intuitive. Similarly, depending the values of the indices m and n, Trhlikova et al.'s (2012) relationship will predict G_{max} approaching either 0 or infinity as p' approaches 0. Hence, such multiplicative forms, in which the effective stress appears as a single parameter or is multiplied to the structure parameter, are unable to reflect the finite G_{max} when $p' = 0$. Chiang and Chae's (1972) relationship for sand has a similar problem when $p' = 0$; see Equation (4.5–1a). For silty clay, Chiang and Chae's (1972) proposed relationship [Equation (4.5–1b)] indicates that G_{max} will approach that of the untreated soil when $p' = 0$. While this allows a finite value of G_{max} to be estimated under unconfined conditions, the G_{max} of the untreated soil may significantly underestimate G_{max} of the treated soil.

TABLE 4.5
Maximum Young's and shear modulus from some studies.

Reference	In situ soil type	Binder type	E_{max} relationship
Chiang and Chae (1972)	Uniform sand and silty clay separated from a glacial outwash	Cement	For uniform sand: $$G_{max} = \left(G_{max\,u} - 0.343A_w p'^{0.5}\right) p'^{0.06A_w} \quad (4.5\text{–}1a)$$ For silty clay: $$G_{max} = G_{max\,u} + A_w(1269 - 2784A_w + 1826A_w^2 - 440A_w^3 + 37A_w^4)p'^{0.45} \quad (4.5\text{–}1b)$$ Here, G_{maxu} is the small-strain shear modulus of the untreated soil and A_w is the cement content in %.
Tanaka and Terashi (1986)	Marine clay	Cement	$G_{max} \sim 107q_u$ to $277q_u$.
Kamruzzaman (2002)	High plasticity marine clay	Cement	$E_{max} = 490q_u$.
Hoyos et al. (2004)	Medium plasticity sulphate-rich clay	(1) Sulphate-resistant type V cement, (2) low calcium class F fly ash and (3) lime mixed with polypropylene fibres	$$G_{max} = Ap'^B \quad (4.5\text{–}2)$$ where p' is the effective confining stress and A and B are fitted parameters.
Puppala et al. (2006)	Medium plasticity sulphate-rich clay	Lime and cement	$$G_{max} = A\ln(t) + B \quad (4.5\text{–}3)$$ Here t is the curing period in hours and A and B are fitted parameters.
Consoli et al. (2010)	Sand	Cement	$$G_{max} = 345q_u\left(\frac{V_v}{V_c}\right)^{0.56} \quad (4.5\text{–}4)$$ where V_v and V_c are the volume of voids and cement, respectively.
Tatsuoka et al. (2010)	Soft clay and sand	Cement	$E_{max} \sim 600q_u$ to $1500q_u$ (ordinary DMM and slurry fill) $E_{max} \sim 400q_u$ (low-strength type DMM)

Verástegui Flores et al. (2010)	Kaolin	OPC and blast furnace cement (CEM III/B)	For OPC-treated kaolin: $$G_{max}/G_{max\,28} = 0.2381\ln(t) + 0.2145 \quad (4.5\text{–}5a)$$ For blast furnace cement-treated kaolin: $$G_{max}/G_{max\,28} = 0.0472t^{1.2866} \quad (\text{for } t < 4 \text{ days}) \quad (4.5\text{–}5b)$$ $$G_{max}/G_{max\,28} = 0.3586\ln(t) - 0.2159 \quad (4.5\text{–}5c)$$ Here, G_{max28} is the small-strain shear modulus after 28 days curing and t is the curing period in days.
Seng and Tanaka (2011)	Fujimori, Kasaoka and Tokyo Bay clays	Cement	$$G_{max} = 310\left(\frac{q_u}{2}\right)^{1.06} \quad (4.5\text{–}6)$$
Trhlikova et al. (2012)	Kaolin	Cement	$$\frac{G_{max}}{p_r} A\left(\frac{p'}{p_r}\right)^n\left(\frac{p_c'}{p'}\right)^m\left(\frac{s}{s_f}\right)^l \quad (4.5\text{–}7)$$ Where p' is the mean effective stress, p_c' is the Hvorslev equivalent stress, $s = p_c'/p_c^{*\prime}$, is a measured of the structure, where $p_c^{*\prime}$ is the Hvorslev equivalent stress of the reconstituted soil, $p_r' = 1$kPa and $s_f = 1$. Parameters A, n, m, and l are fitted parameters.
Yang and Woods (2015)	Mixture of kaolin and bentonite	Type III Portland cement (early high strength)	$$\left(G_{max}/p_a\right) = Ae^{-m}\left(\frac{p'}{p_a} + B(A_w)^2\right)^n \quad (4.5\text{–}8)$$ Here, p_a is the atmospheric pressure, A_w is the cement content expressed as a fraction, e is the post-curing void ratio and A, B, m and n are fitted parameters. For cemented clay, $m = 2.5$. The mean effective stress p' in Equation (4.5–8) includes the effect of soil suction, if any.
Kang et al. (2017)	Dredged clay	Cement	For curing period < 3 days: $$G_{max} = 159q_u^{1.17} \quad (4.5\text{–}9a)$$ For curing period > 3 days: $$G_{max} = 542q_u^{0.89} \quad (4.5\text{–}9b)$$

(Continued)

TABLE 4.5 *(Continued)*
Maximum Young's and shear modulus from some studies.

Reference	In situ soil type	Binder type	E_{max} relationship
Yao et al. (2020)	High plasticity marine clay		(1) For cement-treated clay which has not been subjected to any previous loading episodes, the maximum shear modulus under unconfined condition (i.e. at p' = 0) G_{max0} can be related to the unconfined compressive strength q_u by a relation of the form $G_{max\,0} = Aq_u{}^m$ (4.5–10a) where A and m are fitted parameters. For Singapore marine clay, $A = 1118$ and $m = 0.884$ for G_{max0} and q_u in kPa. (2) For cement-treated clay which has not been loaded past its isotropic primary yield stress[1]: $G_{max} = G_{max\,0} + Bp'^n$ (4.5–10b) where B and n are fitted parameters. For Singapore marine clay, $B = 535$ and $n = 0.952$ for G_{max} and p' in kPa. (2) For cement-treated clay isotropically consolidated normally to a mean effective stress that exceeded its isotropic primary yield stress, the maximum shear modulus under the consolidation pressure $G_{(max)NC}$ is given by $G_{max(NC)} = G_{max\,0}\left(1 - C\Delta e_v\right)$ (4.5–10c) where Δe_v is the void ratio increment from the original unstressed condition. For Singapore marine clay, $C = 3.59$. (3) For cement-treated clay which has been isotropically consolidated to a maximum mean effective stress p'_{max} that is higher than its primary yield stress p'_{py} and then unloaded, the unconfined maximum shear modulus $G_{max0(OC)}$ is given by $G_{max\,0(OC)} = \left\{ C_1 \left(\frac{p'_{max}}{p'_{py}} \right) + C_2 \right\} G_{max\,0}$ (4.5–10d) where C_1 and C_2 are fitted coefficients. For Singapore marine clay, $C_1 = 0.133$ and $C_2 = 0.904$.

[1] The isotropic primary yield stress is defined as the mean effective stress at which non-linearity first appears in the compression curve of void ratio (or specific volume) against mean effective stress.

On the other hand, Yang and Woods (2015) and Yao et al. (2020) proposed relationships in which the effect of effective stress is added to that of structure. This avoids unrealistic estimates of G_{max} when $p' = 0$. Yang and Woods' (2015) relationship, that is Equation (4.5–8), was adopted from Pestana and Salvati's (2006) SimSoil model for maximum shear modulus and shear modulus reduction for uncemented sand. In Equation (4.5–8), the effect of structure is represented by the cement content and void ratio.

Yao et al.'s (2020) model, that is Equation (4.5–10b), can be regarded as a superposition of Seng and Tanaka's (2011) form for the maximum shear modulus under the unconfined condition [Equation (4.5–6)] and Hoyos et al.'s (2004) form [Equation (4.5–2)] for the effect of confining stress. Equation (4.5–10b) was proposed for cement-treated marine clay if the mean effective stress does not exceed the isotropic primary yield stress. The latter is defined as the mean effective stress at which non-linearity first appears in the compression curve of void ratio against the mean effective stress.

Equation (4.5–10a) can be expressed in dimensionally consistent forms as

$$\frac{G_{\max}}{p_a} = A'\left(\frac{q_u}{p_a}\right)^m \tag{4.19}$$

where $A' = \dfrac{A}{p_a^{1-m}}$. For Singapore marine clay, $A' = 669$. This represents the intrinsic component of the shear modulus which arises from the cementation itself.

Similarly, Equation (4.5–10b) can be expressed as

$$\frac{G_{\max}}{p_a} = A'\left(\frac{q_u}{p_a}\right)^m + B'\left(\frac{p'}{p_a}\right)^n \tag{4.20}$$

where $B' = \dfrac{B}{p_a^{1-n}}$. For Singapore marine clay, $B' = 426$. In scenarios where the unconfined compressive strength q_u is significantly higher than p_a (100kPa) and the mean effective stress p' is not much higher than p_a, the intrinsic component, that is $A'\left(\dfrac{q_u}{p_a}\right)^m$, accounts for much of the strength contribution, and the influence of effective stress is of minor significance.

For situations where the mean effective stress exceeds the isotropic primary yield stress, Yao et al. (2020) noted that the rate of increase in maximum shear modulus with respect to effective stress increases. This indicates that the reduction in maximum shear modulus due to loss of cementation caused by yielding is more than compensated by the increase in modulus due to decrease in void ratio, that is densification. Two possible scenarios were considered by Yao et al. (2020). The first is a "normally consolidated" (NC) state, wherein the sample is still subjected to the full isotropic consolidation stress. In such a situation, Equation (4.5–10c) was proposed for estimating the maximum shear modulus $G_{(max)NC}$. The second scenario is an "over-consolidated" (OC) state, wherein the sample was allowed to unload under reduced or zero effective confining stress. For the special case, where the sample was unloaded to an unconfined state, Equation (4.5–10d) was proposed for estimating the unconfined maximum shear modulus of the over-consolidated sample. The effect of stress history [Equation (4.5–10d)] is represented by an "overload" ratio of the maximum mean effective stress to the primary yield stress.

In most practical situations, the mean effective stress encountered is often lower than the primary yield stress since the objective of hardening the soil is to limit strain and ground movement. In such cases, consideration of densification and stress history is unlikely to be required, and Equations (4.19) and (4.20) will suffice. Furthermore, in scenarios where the unconfined compressive strength q_u is significantly higher than p_a (100kPa) and the mean effective stress p' is not much higher than p_a, the intrinsic component, that is $A'\left(\frac{q_u}{p_a}\right)^m$, accounts for much of the strength contribution, and the influence of effective stress is of minor significance. In such cases, Equation (4.19) will suffice.

4.4.2 Small-Strain Non-Linearity

Compared to the maximum shear modulus, fewer studies have been conducted into the degradation of shear modulus with shear strain. Based on resonant column results, Yang and Woods (2015) proposed that the change in tangent shear modulus G_{tan} of cement-treated clay can be represented by

$$\frac{G_{tan}}{G_{max}} = \frac{1}{1+\omega_c \zeta_s^{0.75} + \omega_c \zeta_s + \omega_a^2 \zeta_s^2} \tag{4.21}$$

where ζ_s is a dimensionless stress measure that is defined as a function shear and normal stresses and ω_a is a fitted parameter. The parameter ω_c is given by

$$\omega_c = \omega_s + e\sqrt{B\left(A_w^2\right)} \tag{4.22}$$

where e is the void ratio, ω_s is a fitted parameter and B is the same fitted parameter as in Equation (4.5–8). The variable ζ_s is a normalised shear stress, defined as

$$\zeta_s = \frac{\sqrt{2}\left(\tau/p_a\right)}{\left(\sigma/p_a\right)+B\left(A_w^2\right)} - \frac{\sqrt{2}\left(\tau_{rev}/p_a\right)}{\left(\sigma_{rev}/p_a\right)+B\left(A_w^2\right)} \tag{4.23}$$

where (σ, τ) are the current shear and normal stresses, p_a is the atmospheric pressure and $(\sigma_{rev}, \tau_{rev})$ are the normal and shear stresses at the most recent stress reversal (transition between loading and unloading) point.

Equation (4.21) is a relation between the tangent shear modulus and the shear stress. This differs from the conventional form of modulus degradation relation between secant shear modulus G_{sec} and shear strain. Since $G_{tan} = \frac{d\tau}{d\gamma}$ and $G_{sec} = \frac{\tau}{\gamma}$, the conversion from G_{tan} to G_{sec} involves an integration operation.

Using resonant column and local strain data from triaxial tests, as well as G_{max} measured by bender element tests, Yao et al. (2018) proposed empirical relationships between G_{sec} and generalised shear or deviatoric strain ε_s, the latter being defined by

$$\varepsilon_s = \frac{2}{3}\sqrt{\frac{1}{2}\left\{\left(\varepsilon_x - \varepsilon_y\right)^2 + \left(\varepsilon_y - \varepsilon_z\right)^2 + \left(\varepsilon_z - \varepsilon_x\right)^2 + \frac{3}{4}\left(\gamma_{xy}^2 + \gamma_{yz}^2 + \gamma_{zx}^2\right)\right\}} \quad (4.24)$$

where ε_x, ε_y and ε_z are the normal strain components and γ_{xy}, γ_{yz} and γ_{zx} are the shear strain components in three mutually orthogonal x-, y- and z-directions. ε_s can be related to the torsional shear strain generated by resonant column through

$$\varepsilon_s = \frac{\gamma}{\sqrt{3}} \quad (4.25)$$

For unconfined conditions, Yao et al. (2018) proposed a modified hyperbolic relationship between G_{sec} and ε_s, given by

$$\frac{G_{sec}}{G_{max}} = \frac{1 - \left(G_f / G_{max}\right)}{1 + \left(\varepsilon_s / \varepsilon_r\right)^{\alpha}} + \frac{G_f}{G_{max}} \quad (4.26)$$

where G_f is a "degraded" shear modulus reached at large strain and ε_r and α are model parameters. The parameter ε_r represents the reference strain level at which $G_{sec} = 0.5G_{max}$. It can be estimated directly from the test data and do not need to be fitted. The parameter α can be readily fitted by expressing Equation (4.26) in the form

$$\frac{G_{max} - G_{sec}}{G_{sec} - G_f} = \left(\frac{\varepsilon}{\varepsilon_r}\right)^{\alpha} \quad (4.27)$$

By plotting $\frac{G_{max} - G_{sec}}{G_{sec} - G_f}$ against $\left(\frac{\varepsilon}{\varepsilon_r}\right)$, α can be determined using a power-law fit. Alternatively, α can also be determined through a linear fit of $\log\left(\frac{G_{max} - G_{sec}}{G_{sec} - G_f}\right)$ against $\log\left(\frac{\varepsilon}{\varepsilon_r}\right)$. Yao et al. (2018) also proposed empirical relationships for G_f / G_{max}, ε_r and α against unconfined compressive strength. As Figure 4.7 shows, as q_u increases, the downturn in the degradation curve starts earlier, indicating a smaller range of shear strain with true elastic behaviour. This is reflected in the decrease in reference strain ε_r. On the other hand, the gradient of the curve at the downturn segment is smaller, indicating a slower rate of decrease, which is also reflected in the smaller value of α.

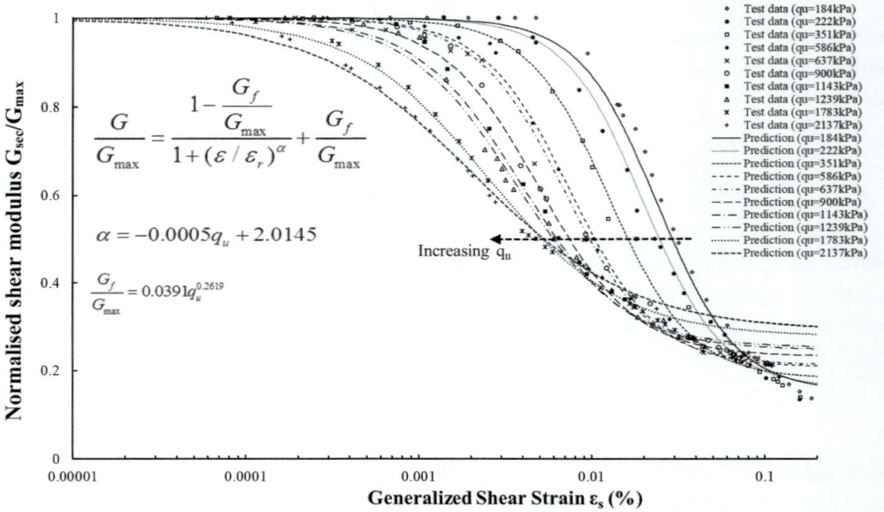

FIGURE 4.7 Experimental data and fitted modulus degradation curves for different unconfined compressive strength.

Source: **Yao et al. (2018).**

For confined conditions, Yao et al. (2018) showed that the modulus degradation curves can still be described using Equation (4.26), but with different values of ${G_f}/{G_{max}}$, ε_r and α. Yao et al. (2018) also proposed equations relating these three parameters to the mean effective stress. In general, as the mean effective stress increases, all three parameters increase. The increase in ε_r indicates a later onset of the downturn whereas the increase in α indicates a higher gradient of the downturn portion of the degradation curve.

There are some differences between Yang and Woods' (2015) and Yao et al.'s (2018) studies. Yang and Woods (2015) used cement content ranging from 2.5% to 7.5%, whereas Yao et al. (2018) used cement content ranging from 20% to 50%. In Yang and Woods' (2015) study, the untreated clay consists of kaolinite, bentonite or a mixture of the two. Yao et al. (2018) used a natural marine clay, which is a high-plasticity inorganic clay. Yang and Woods (2015) reported no change in maximum shear modulus with confining pressure, whereas Yao et al. (2020) reported an increase in maximum shear modulus with confining pressure. Yao et al. (2018), using local strain triaxial measurements, were able to reach generalised shear strain levels of about 5%, this being higher than Yang and Woods' (2015) maximum shear strain of less than 1%. However, Yang and Woods (2015) measured damping ratio whereas Yao et al. (2018) did not. Yao et al.'s (2018) degradation curve requires one more parameter than Yang and Woods' (2015), but the parameters are relatively easy to evaluate. Evaluation of ω_a and ω_s in Equations (4.21) and (4.22) may, on the other hand, require some trial-and-error to evaluate.

4.5 YIELDING AND CONSTITUTIVE MODEL

The constitutive behaviour of cement-treated soils depends upon the soil that is treated. In particular, the behaviour of cement-treated sand differs significantly from that of cement-treated clay. In general, the void ratio of cement-treated intermediate and high plasticity clay at and after mixing is much higher than that of cement-treated sand. Table 4.6 shows the water content and void ratio of cement-treated sand and clay from some studies. As discussed in Chapter 3, the typical range of water content for cement–soil mixing is about 1.3 to 1.5 times the liquid limit of the untreated soil. For this reason, more water is often required to mix cement with high plasticity clay than with sand. This results in cement-treated clay often having higher water content and void ratio than cement-treated sand. These differences have significant implication on the constitutive behaviour of the treated soil.

In general, cement-treated sand tested in triaxial condition is usually strongly dilatant post-yielding for effective confining stress up to about 500kPa, even for relatively loose-packed sand particles (e.g. Ajorloo et al. 2012; Amini and Hamidi 2014; Wang et al. 2021). On the other hand, cement-treated clays (e.g. Chiu et al. 2008; Xiao et al. 2014) often exhibit much larger plastic compressive volumetric response under isotropic compression and low stress ratio. The term "stress ratio" is used herein to refer to the ratio of deviator stress (q) to mean effective stress (p').

This is reflected in constitutive models which have been proposed for cement-treated soils. For instance, Namikawa and Mihira's (2007) model for cement-treated sand comprises linear yield and failure envelopes for shear and tensile loading (Figure 4.8). The tensile yield and failure envelopes describe yielding and failure at low and negative mean effective stresses. However, yield surfaces were not proposed yielding under isotropic and low $\left(q/p' \right)$ stress ratios. Prior to initial yield,

TABLE 4.6
Some engineering properties of cement-treated sand and high-plasticity clay.

Parameter	Cement-treated sand	Cement-treated clay
Water content (%)	• ~9% to ~18% at mixing (Ajorloo et al. 2012). • ~10% post-curing (Gu et al. 2022).	• ~70% to ~88% at mixing. Based on Singapore industry deep-mixing norm of using 180kg cement per cubic metre of ground for high plasticity Singapore marine clay. • ~71% to ~183% at mixing (Xiao et al. 2014). For high plasticity inorganic marine clay.
Void ratio	• 0.37 to 0.42 after isotropic consolidation (Amini and Hamidi 2014). • 0.32 to 0.65 at start of isotropic consolidation (Dong et al. 2023).	• ~2.0 to ~2.4 at mixing. Based on Singapore industry deep-mixing norm of using 180kg cement per cubic metre of ground for high plasticity Singapore marine clay. • ~1.5 to 2.5 post-curing (Chiu et al. 2008). For intermediate plasticity dredged freshwater and marine clay.

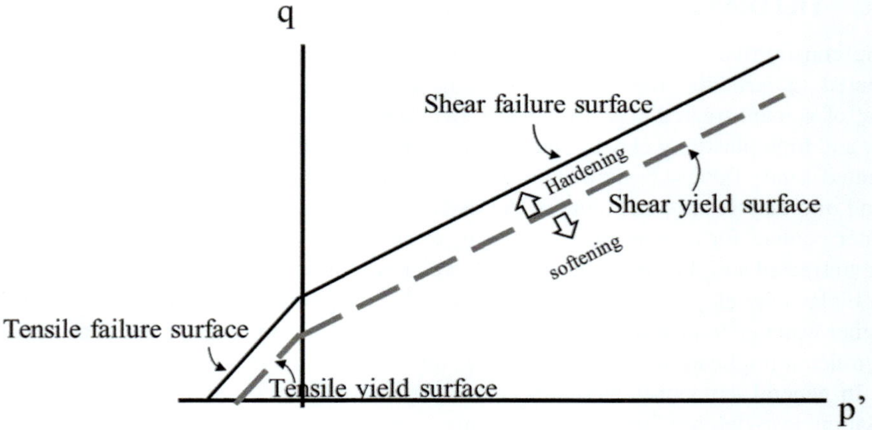

FIGURE 4.8 Yield and failure surfaces in Namikawa and Mihira's (2007) model for cement-treated sand.

the shear yield surface lies below the shear failure surface whereas the tensile yield surface is coincident with tensile failure envelope. Initial yielding on the shear yield surface leads to hardening with the yield surface migrating towards the shear failure surface. Post-failure behaviour is characterised by softening, with the tensile and yield surfaces falling back below the failure surfaces.

Gao and Zhao (2012) also used a Mohr-Coulomb type yield surface, to describe the yielding of cement-treated sand. A cross-anisotropic fabric tensor was also incorporated to account for fabric-induced anisotropy.

Abdulla and Kiousis (1997) modelled the behaviour of cement-treated sand using a superposition approach on a "mixture" of sand and cement phases. The sand phase is modelled using a non-associated Drucker-Prager model while the cement phase is assumed to behave as a hardening, non-associated von-Mises model. Hardening of the cement phase is modelled by increasing the von-Mises yield stress. Volumetric compression of the cement phase is modelled through a plastic potential G_c of the form

$$G_c = q - \beta p' \tag{4.28}$$

where β is a compression coefficient. However, at low sand content, much of the volumetric compression from the cement phase is likely to be offset by the dilation of the sand phase.

A common approach for modelling the behaviour naturally and lightly of cement clay is by expanding the elliptical Cam Clay yield surface of the uncemented soil into a larger structure surface for the cemented soil while keeping the shape unchanged. An example of this is the Sydney soil model (Liu et al. 2011, 2012; Airey et al. 2011), which was proposed for naturally cemented clay. In this model, the structured yield surface is obtained by incorporating a void ratio increment to the unstructured yield surface. As the structured yield surface passes through the origin in stress space, true cohesion is not considered.

In some models, true cohesion is modelled by left-shifting the structure surface so that the yield locus has a cohesion intercept (e.g. Nova et al. 2003; Lee et al. 2004). Loss of structure is commonly assumed to be related to plastic shear strain, but not plastic volumetric strain (Lee et al. 2004; Suebsuk et al. 2010). Arroyo et al. (2012) also used a similar concept. This allows the yield locus to be left-shifted while preserving the same shape. Arroyo et al.'s (2012) model was verified with cement-treated clay specimens with up to 10% cement content. Loss of structure was assumed to be a function of maximum mean effective stress experienced by the soil, that is the isotropic pre-consolidation stress. This implicitly assumes that loss of structure is the result of volumetric strain. Their model was validated with cement-treated clay specimens up to 10% cement content under undrained triaxial shearing.

However, cement-treated clay, especially at high cement content, have been noted to have a yield surface which differs significantly from an elliptical shape (e.g. Xiao et al. 2014). Hence, some modification to the shape of the Cam Clay yield surface is also employed. For instance, Kasama et al. (2000) used a modification of the Cam Clay work equation to alter the shape of the yield surface to model the behaviour cement-treated clay for cement content up to 5%. Cohesion was modelled using an added mean effective stress (p') concept, which is equivalent left-shifting the yield surface into the negative effective stress regime. However, loss of cementation, hardening and softening behaviour was not prescribed. Horpibulsuk et al. (2010) and Suebsuk et al. (2010) also used a similar approach to model the behaviour of cement-treated clay, for cement content up to 18%.

Another approach superimposes the behaviour of the uncemented soil and that of the cementation (Gens et al. 2005; Liu and Shen 2005) to describe the constitutive behaviour of the cement-treated soil. This approach is similar to that of Abdulla and Kiousis (1997) for cement-treated sand. A drawback with this approach is the large number of parameters required, most of which need to be determined by fitting. Moreover, cement-treated clay is mineralogically altered from the untreated clay (Chew et al. 2004). Hence, it is difficult to see how direct superposition can be applicable.

Xiao et al. (2017) proposed a bounding surface model based on incorporating cohesion into the work equation for Cam Clay. This leads to a non-elliptical initial yield surface, which transforms into an elliptical shape as cohesion decreases with plastic strain. The size of the initial yield surface is determined by a sensitivity parameter, which determines the initial isotropic yield stress (Xiao and Lee 2014). Both cohesion and sensitivity can be related to the unconfined compressive strength and can thus be readily and independently determined. The sensitivity parameter also decreases with work associated with plastic strains. The rates of deterioration of the cohesion and sensitivity parameters are evaluated empirically from isotropic compression and triaxial test data. The initial yield surface as well as the deterioration in the cohesion and sensitivity parameters, which describe the structure (including cementation) of the soil, is energy based. This has the advantage of ensuring that energy is continually being dissipated, rather than generated, during plastic straining. Xiao et al.'s (2017) cohesive Cam Clay model was verified for cement content between 10% and 50%.

This discussion is by no means exhaustive and new models will no doubt continue to be developed. Nonetheless, regardless of whether one is using an existing model or developing a new model, there are some aspects of cement-treated clay behaviour which are relevant. These are as follows:

1. Enhanced initial isotropic yield stress, compared to the uncemented soil, which is typically assumed to be related to the size of the initial yield surface.
2. Non-zero deviator stress (q) corresponding to zero mean effective stress (p'), owing to true cohesion. This is reflected as a non-zero q-intercept between the yield surface and the q-axis. Cohesion may not be a significant component at very low cement content but is likely to be more significant as cement content increases. This may be important in soil stabilisation for underground construction such as deep excavation and tunnelling works, where cement content of 20% or higher is often used.
3. Decrease in cementation and structure parameters to reflect loss of cementation and structure as straining takes place. This is a departure from the conventional Mohr-Coulomb model, where cohesion is often assumed to be a constant.

Finally, from a practical viewpoint, the constitutive model should also be economical in terms of the number of parameters. Too many parameters will increase the effort, as well as uncertainty, in terms of parametric evaluation.

REFERENCES

Abdulla, A. A. and Kiousis, P. D. (1997). Behavior of cemented sands—II: Modelling. *International Journal for Numerical and Analytical Methods in Geomechanics*, 21: 549–568.

Abrams, D. A. (1919). Design of concrete mixtures. In *Bulletin, structural materials research laboratory*, Lewis Institute, Chicago, Bulletin 1.

Airey, D., Carter, J. and Liu, M. (2011). Sydney soil model. II: Experimental validation. *International Journal of Geomechanics*: 225–238. https://doi.org/10.1061/(ASCE)GM.1943-5622.0000079.

Ajorloo, A. M., Mroueh, H. and Lancelot, L. (2012). Experimental investigation of cement treated sand behaviour under triaxial test. *Geotechnical and Geological Engineering*, 30: 129–143.

Amini, Y. and Hamidi, A. (2014). Triaxial shear behaviour of a cement-treated-sand-gravel mixture. *Journal of Rock Mechanics and Geotechnical Engineering*, 6: 455–465.

Arroyo, M., Ciantia, M., Castellanza, R., Gens, A. and Nova, R. (2012). Simulation of cement-improved clay structures with a bonded elastoplastic model: A practical approach. *Computers and Geotechnics*, 45: 140–150.

Asano, J., Ban, K., Asuma, K. and Takahashi, K. (1996). Deep mixing method of soil stabilization using coal ash. *Proc. Is-Tokyo '96 - 2nd Int. Conf. Ground Improvement Geosystems* (Eds. R. Yonekura, M. Terashi, and M. Shibazaki), 14–17 May: 393–398.

Asghari, E., Toll, D. G. and Haeri, S. M. (2003). Triaxial behaviour of a cemented gravely sand, Tehran alluvium. *Geotechnical and Geological Engineering*, 21: 1–28.

ASTM (2011). *ASTM D2487–11: Standard practice for classification of soils for engineering purposes (Unified soil classification system)*, ASTM International, West Conshohocken, PA.
Chen, J., Liu, Y. and Lee, F. H. (2016). A statistical model for the unconfined compressive strength of deep-mixed columns. *Géotechnique*, 66(5): 351–365.
Chew, S. H., Kamruzzaman, A. H. M. and Lee, F. H. (2004). Physicochemical and engineering behaviour of cement treated clays. *Journal of Geotechnical and Geoenvironmental Engineering*, 130(7): 696–706.
Chian, S. C., Nguyen, S. T. and Phoon, K. K. (2016). Extended strength development model of cement-treated clay. *Journal of Geotechnical and Geoenvironmental Engineering (ASCE)*, 142(2): 06015014.
Chiang, Y. C. and Chae, Y. S. (1972). Dynamic properties of cement-treated soils. *Highway Research Record*, 379: 39–51.
Chin, K. G. (2006). *Constitutive behaviour of cement treated marine clay*. PhD thesis, National University of Singapore.
Chiu, C. F., Zhu, W. and Zhang, C. L. (2008). Yielding and shear behaviour of cement-treated dredged materials. *Engineering Geology*, 103: 1–12.
Consoli, N. C. (2014). A method proposed for the assessment of failure envelopes of cemented sandy soils. *Engineering Geology*, 169: 61–68.
Consoli, N. C., Cruz, R. C., Consoli, B. S. and Maghous, S. (2012). Failure envelope of artificially cemented sand. *Geotechnique*, 62(6): 543–547.
Consoli, N. C., Cruz, R. C., Floss, M. F. and Festugato, L. (2010). Fundamental parameters controlling tensile and compressive strength of artificially cemented sand. *Journal of Geotechnical and Geoenvironmental Engineering*, 136(5): 759–763.
Consoli, N. C., Foppa, D., Festugato, L. and Heineck, K. S. (2007). Key parameters for strength control of artificially cemented soils. *Journal of Geotechnical and Geoenvironmental Engineering*, 133(2): 197–205.
Consoli, N. C., Samaniego, R. A. Q., Marques, S. F. V., Venson, G. I., Pasche, E. and Velasquez, L. E. G. (2016). A single model establishing strength of dispersive clay treated with distinct binders. *Canadian Geotechnical Journal*, 53(12): 2072–2079.
Constantinides, G. and Ulm, F.-J. (2004). The effect of two types of C-S-H on the elasticity of cement-based materials: Results from nanoindentation and micromechanical modelling. *Cement and Concrete Research*, 34: 67–80.
Dong, X., Bao, X., Cui, H., Xu, C. and Chen, X. (2023). Isotropic compression and triaxial shear behaviors of cement- and cement-gravel-treated granite residual soil for use as subgrade filling. *Construction and Building Materials*, 390: 131780.
EN ISO 14688–2 (2018). *EN ISO 14688–2:2018: Geotechnical investigation and testing—Identification and classification of soil—Part 2: Principles for a classification*. Comité Européen de Normalisation, Brussels.
Fernandez, R., Martirena, F. and Scrivener, K. L. (2011). The origin of the pozzolanic activity of calcined clay minerals: A comparison between kaolinite, illite and montmorillonite. *Cement and Concrete Research*, 41(1): 113–122.
Futaki, M., Nakano, K. and Hagino, Y. (1996). Design strength of soil-cement columns as foundation ground for structures. *Proc. 2nd Int. Conf. on Ground Imp. Geosystems: Grouting and Deep Mixing*, Balkema: 481–484.
Gallavresi, F. (1992). Grouting improvement of foundation soils. *Proc. Grouting, Soil Improvement and Geosynthetics*, ASCE, New York, 1: 1–38.
Gao, Z. and Zhao, J. (2012). Constitutive modeling of artificially cemented sand by considering fabric anisotropy. *Computers & Geosciences*, 41: 57–69.

Gens, A., Vaunat, J. and Garitte, B. (2005). Elastoplastic modeling of hard soils and soft rocks: Formulation and application. Proc., VIII Int. Conf. on Computational Plasticity COMPLAS VIII (Eds. E. Oriate and D. R. J. Owen), CIMNE, Barcelona, Spain: 1–4.

GeoTesting Express (1996). Geotechnical tests on soil-cement mix for Central Artery/Tunnel Project. *Final Report to Bechtel/Parsons Brinckerhoff*, GeoTesting Express.

Gu, J., Lyu, H., Yang, J. and Zeng, C. (2022). Effects of cement content and curing period on geotechnical properties of cement-treated calcareous sands. *Transportation Geotechnics*, 33: 100732.

Guo, X., Zhang, W., Yu, H. and Jin, D. (2022). Reduction, stabilization, and solidification of Cr(VI) in contaminated soils with a sustainable by-product-based binder. *Chemosphere*, 307(2): 135902.

Hayashi, H., Nishikawa, J., Ohishi, K. and Terashi, M. (2003). Field observation of long-term strength of cement treated soil. *Proceedings of the Third International Conference on Grouting and Ground Treatment*, 1: 598–609.

Horpibulsuk, S., Liu, M. D., Liyanapathirana, D. S. and Suebsuk, J. (2010). Behaviour of cemented clay simulated via the theoretical framework of structured cam clay model. *Computers and Geotechnics*, 37(1–2): 1–9.

Horpibulsuk, S., Miura, N. and Nagaraj, T. S. (2003). Assessment of strength development in cement-admixed high water content clays with Abrams' law as a basis. *Geotechnique*, 53(4): 439–444.

Horpibulsuk, S., Rachan, R., Suddeepong, A. and Chinkulkijniwat, A. (2011). Strength development in cement admixed Bangkok clay: Laboratory and field investigations. *Soils and Foundations*, 51(2): 239–251.

Hoyos, L. R., Puppala, A. J. and Chainuwat, P. (2004). Dynamic properties of chemically stabilized sulfate rich clay. *Journal of Geotechnical and Geoenvironmental Engineering*, 130(2): 153–162.

Ichiba, T., Ikegami, M., Masuda, K., Tsuruya, H. and Satou, S. (2002). Physical properties and strength of cement-treated marine clay after 20 years. *Proc. Ann. Conf. Japan Soc. Civ. Eng.*: 121–122 (in Japanese).

Ikegami, M., Ichiba, T., Ohishi, K. and Terashi, M. (2006). Long-term properties of cement treated soil 20 years after construction. *Proc. 16th Int. Conf. Soil Mech. Geotech. Eng.: Geotechnology in Harmony with the Global Environment, Osaka, Japan*: 1199–1202. 12–15 September, ISBN-13 978-9059660274

Japan Cement Association (2007). *Ground improvement manual using cement-based solidification material*. 3: 43–44.

Kamruzzaman, A. H. M. (2002). *Physico-chemical and engineering behavior of cement treated Singapore marine clay*. PhD dissertation, National University of Singapore.

Kang, G., Tsuchida, T. and Kim, Y. (2017). Strength and stiffness of cement-treated marine dredged clay at various curing stages. *Construction and Building Materials*, 132(2017): 71–84.

Kasama, K., Ochiai, H. and Yasufuku, N. (2000). On the stress-strain behaviour of lightly cemented clay based on an extended critical state concept. *Soils and Foundations*, 40(5): 37–47.

Kawasaki, T., Niina, A., Saitoh, S. and Suzuki, Y. (1981). Deep Soil Mixing. *Proc. 10th Int. Conf. on Soil Mech. Found. Engrg*, Stockholm: 721–724. Balkema.

Kida, T., Kubo, H. and Sumida, M. (1977). Study on disposal and treatment of mud (Part 3). *Proc. 12th Annual Meeting of Japanese Soc. Soil Mech. Found. Engg.*: 1309–1312 (In Japanese).

Kitazume, M., Nakamura, T., Terashi, M. and Ohishi, K. (2003). Laboratory tests on long-term strength of cement treated soil. *Proceedings of the Third International Conference on Grouting and Ground Treatment*, 1: 586–597.

Kitazume, M. and Takahashi, H. (2009). 27 years' investigation on property of in-situ quicklime treated clay. *Proc. 17th Int. Conf. Soil Mech. Geotech. Eng., Alexandria, Egypt, 5–9 October.* IOS Press: 2358–2361.

Koseki, J., Sato, T., Mihira, S., Takeya, N. and Yoshizawa, M. (2005). Comparison of tensile strength of cement treated sand by various test methods. *Deep Mixing 05*: 95–100.

Kulkov, S. N., Buyakova, S. P. and Maslovskii, V. I. (2003). Structure, phase composition and mechanical properties of zirconia-based ceramics. *Vestnik Tomskogo Gosudarstvennogo Universiteta. Seriya Fizika (Bulletin of Tomsk State University, Physics Series)*, 13(34): 34–57 (in Russian).

Lee, F. H., Lee, Y., Chew, S. H. and Yong, K. Y. (2005). Strength and modulus of marine clay-cement mixes. *Journal of Geotechnical and Geoenvironmental Engineering*, 131(2): 178–186.

Lee, K., Chan, D. and Lam, K. (2004). Constitutive model for cement treated clay in a critical state frame work. *Soils and Foundations*, 44(3): 69–77.

Lim, S. M., He, L., Goh, S. H. and Lee, F. H. (2021). Strength of chemically stabilized sewage sludge—some inferences from recent studies. *Geotechnics*, 1(2): 573–587.

Liu, E. L. and Shen, Z. J. (2005). Binary medium model for structured soils. *Journal of Hydraulic Engineering*, 36(4): 391–395 (in Chinese).

Liu, J., Yang, K. and Gurpersaud, N. (2022). Tensile strength of cement-treated Champlain Sea clay. *Geotechnical and Geological Engineering*, 40: 5467–5480.

Liu, M. D., Carter, J. P. and Airey, D. W. (2011). Sydney soil model. I: Theoretical formulation. *Int. Jnr. Geomech.*, 11 (3): 211–224.

Liu, M. D., Carter, J. P. and Airey, D. W. (2012). Introduction to the Sydney Soil Model. *Proc. Int. Symp. Sustainable Geosynthetics and Green Tech. for Climate Change (SGCC2011)*, 20–21 June: 409–428.

Locat, J., Tremblay, H. and Leroueil, S. (1996). Mechanical and hydraulic behaviour of a soft inorganic clay treated with lime. *Canadian Geotechnical Journal*, 33: 654–669.

Muhunthan, B. and Sariosseiri, F. (2008). *Interpretation of geotechnical properties of cement treated soils*. WSDOT Res. Rep. WA-RD 715.1, Washington State Department of Transportation.

Nakamura, T. and Kitazume, M. (2006). *Laboratory tests on durability of cement treated soils*. Tech. Note Port and Airport Res. Inst.: 1128, 13p (in Japanese).

Nakarai, K. and Yoshida, T. (2015). Effect of carbonation on strength development of cement-treated Toyoura silica sand. *Soils and Foundations*, 55(4): 857–865.

Namikawa, T. and Koseki, J. (2007). Evaluation of tensile strength of cement-treated sand based on several types of laboratory tests. *Soils and Foundations*, 47(4): 657–674.

Namikawa, T. and Mihira, S. (2007). Elasto-plastic model for cement-treated sand. *International Journal for Numerical and Analytical Methods in Geomechanics*, 31: 71–107.

Nova, R., Castellanza, R. and Tamagnini, C. (2003). A constitutive model for bonded geomaterials subject to mechanical and/or chemical degradation. *International Journal for Numerical and Analytical Methods in Geomechanics*, 27(9): 705–732.

Ohishi, K., Ikegami, M., Masuda, K., Ichiba, T. and Tsuruya, H. (2002). Long-term durability of cement-treated marine clay after 20 years. *Proceedings of the Annual Conference of the Japan Society of Civil Engineers*: 123–124 (in Japanese).

Pan, Y. T., Xiao, H. W., Lee, F. H. and Phoon, K. K. (2016). Modified isotropic compression relationship for cement-admixed marine clay at low confining stress. *Geotechnical Testing Journal*, 39(4): 1–8.

Pellenq, R. J.-M., Kushima, A., Shahsavari, R., Van Vliet, K. J., Buehler, M. J., Yip, S. and Ulm, F.-J. (2009). A realistic molecular model of cement hydrates. *Proceedings of the National Academy of Sciences of the United States of America*, 106(38): 16102–16107.

Pestana, J. M. and Salvati, L. A. (2006). Small-strain behavior of granular soils. Part 1: Model for cemented and uncemented sands and gravels. *Journal of Geotechnical and Geoenvironmental Engineering*, 132(8): 1071–1081. https://doi.org/10.1061/(ASCE)1090-0241(2006)132:8(1071).

Powers, T. C. (1958). Structure and properties of hardened Portland cement paste. *Journal of the American Ceramic Society*, 41: 1.

Puppala, A. J., Kadam, R., Madhyannapu, R. S. and Hoyos, L. R. (2006). Small-strain shear moduli of chemically stabilized sulfate-bearing cohesive soils. *Journal of Geotechnical and Geoenvironmental Engineering*, 132(3): 322–336.

Rao, S. N. and Rajasekaran, G. (1996). Reaction products formed in lime-stabilized marine clays. *Journal of Geotechnical Engineering*, 122(5): 329–336.

Rios, S., da Fonseca, A. V. and Baudet, B. A. (2014). On the shearing behaviour of an artificially cemented soil. *Acta Geotechnica*, 9: 215–226.

Saitoh, S., Kawasaki, T., Nina, S., Babasaki, R. and Miyata, T. (1980). Research on DMM using cementitious agents (part 10)-engineering properties of treated soils. *Proc. 15th Nat. Conf. JSSMFE*, Tokyo: 717–720 (in Japanese).

Saitoh, S., Suzuki, Y., Nishioka, S. and Okumura, R. (1996). Required strength of cement improved ground. *Proc. 2nd Int. Conf. on Ground Imp. Geosystems: Grouting and Deep Mixing*, Balkema: 557–562.

Seng, S. and Tanaka, H. (2011). Properties of cement-treated soils during initial curing stages. *Soils and Foundations*, 51(5): 775–784.

Shahrin, R. and Bobko, C. P. (2017). Characterizing strength and failure of calcium silicate hydrate aggregates in cement paste under micropillar compression. *Journal of Nanomechanics and Micromechanics (ASCE)*, 7(4): 06017002.

Shooshpasha, I. and Shirvani, R. A. (2015). Effect of cement stabilization on geotechnical properties of sandy soils. *Geomechanics and Engineering*, 8(1): 17–31.

Subramaniam, S., Moon, S. W., Moon, J. and Ku, T. (2018). CSA-treated sand for geotechnical application: Microstructure analysis and rapid strength development. *Journal of Materials in Civil Engineering*, 30(12): 04018313.

Suebsuk, J., Horpibulsuk, S. and Liu, M. D. (2010). Modified structured cam clay: A generalised critical state model for destructured, naturally structured and artificially structured clays. *Computers and Geotechnics*, 37: 956–968.

Tanaka, H. and Terashi, M. (1986). Properties of treated soil formed in situ by deep mixing method. *Reports of the Port and Harbour Research Institute*, 25(2): 89–119 (in Japanese).

Tang, Y. X., Miyazaki, Y. and Tsuchida, T. (2001). Practices of reused dredging by cement treatment. *Soils and Foundations*, 41(5): 129–143.

Tariq, T. A. and Maki, T. (2014). Mechanical behaviour of cement-treated sand. *Construction and Building Materials*, 58: 54–63.

Tatsuoka, F. (2010). Cement-mixed soils in Trans-Tokyo Bay Highway Project. *Soils and Foundations*, 50(6): 785–804.

Terashi, M., Okumura, T. and Mitsumoto, T. (1977). Fundamental properties of lime and cement treated soils (1st Rep.). *Reports of the Port and Harbour Research Institute*, 16(1): 3–28 (in Japanese).

Terashi, M., Tanaka, H., Mitsumoto, T., Honma, S. and Ohhashi, T. (1983). Fundamental properties of lime and cement treated soils (3rd Rep.). *Reports of the Port and Harbour Research Institute*, 22(1) (in Japanese).

Terashi, M., Tanaka, H., Mitsumoto, T., Niidome, Y. and Honma, S. (1980). Fundamental properties of lime and cement treated soils (2nd Rep.). *Reports of the Port and Harbour Research Institute*, 19(1): 33–62 (in Japanese).

Trhlikova, J., Masin, D. and Bohac, J. (2012). Small-strain behaviour of cemented soils. *Geotechnique*, 62(10): 943–947.

Tsuchida, T. and Tang, Y. X. (2015). Estimation of compressive strength of cement-treated marine clays with different initial water contents. *Soils and Foundations*, 55(2): 359–374.

Uddin, K., Balasubramaniam, A. S. and Bergado, D. T. (1997). Engineering behavior of cement-treated Bangkok soft clay. *Geotechnical Engineering*, 28(1): 89–119.

Verástegui Flores, R. D., Di Emidio, G. and Van Impe, W. F. (2010). Small-strain shear modulus and strength increase of cement-treated clay. *Geotechnical Testing Journal*, 33(1): 62–71. Paper ID GTJ102354.

Viggiani, G. and Atkinson, J. H. (1995). Stiffness of fine-grained soil at very small strains. *Géotechnique*, 45(2): 249–265.

Wang, L., Chu, J., Wu, S. and Wang, H. (2021). Stress–dilatancy behavior of cemented sand: Comparison between bonding provided by cement and biocement. *Acta Geotechnica*, 16: 1441–1456.

Xiao, H. W. and Lee, F. H. (2014). An energy-based isotropic compression relation for cement-admixed soft clay. *Geotechnique*, 64(5): 412–418.

Xiao, H. W., Lee, F. H. and Chin, K. G. (2014). Yielding of cement-treated marine clay. *Soils and Foundations*, 54(3): 488–501.

Xiao, H. W., Lee, F. H. and Liu, Y. (2017). Bounding surface Cam-Clay model with cohesion for cement-admixed clay. *International Journal of Geomechanics*, 17(1): 04016026.

Yamashita, E., Cikmit, A. A., Tsuchida, T. and Hashimoto, R. (2020). Strength estimation of cement-treated marine clay with wide ranges of sand and initial water contents. *Soils and Foundations*, 60: 1065–1083.

Yang, L. and Woods, R. D. (2015). Shear stiffness modeling of cemented clay. *Canadian Geotechnical Journal*, 52: 156–166.

Yao, K., Chen, Q., Ho, J., Xiao, H. and Lee, F. H. (2018). Strain-dependent shear stiffness of cement-treated marine clay. *Journal of Materials in Civil Engineering*, 30(10): 04018255.

Yao, K., Chen, Q., Xiao, H., Liu, Y. and Lee, F. H. (2020). Small-strain shear modulus of cement-treated marine clay. *Journal of Materials in Civil Engineering*, 32(6): 04020114. https://doi.org/10.1061/(ASCE)MT.1943-5533.0003153.

5 Quality Control

Fook-Hou Lee and Yong Liu

5.1 TREATED SOIL PARAMETERS FOR QUALITY ASSESSMENT

The parameters which are assessed for acceptability in chemical treatment of soil depends primarily on the aim of the improvement process. For instance, if the objective of improvement is to stabilise and solidify contaminated soil and prevent leakage of toxic chemicals such as heavy metals and hydrocarbons, then hazardous leachate contents will be an important parameter (e.g. Conner 1994). If the objective is to provide seepage cut-off, such as in cut-off walls, then (low) permeability and continuity of the cut-off structure are important parameters. However, if the objective is to stabilise soft and marginal soils for construction, then the important parameters are strength, stiffness and spatial uniformity. An important aspect of chemical soil mixing is spatial variability. Indeed, this is an important aspect which makes adequate monitoring all the more important. This issue will be discussed in more detail in this as well as in the next chapter.

In a typical construction setting in which soil improvement is required, the design engineer of the permanent structure, such as foundations, or temporary construction, such as earth retaining structures for deep excavations, would determine a design strength of the treated soil that is required to ensure stability and limit ground movement. This leads to the specification of a "criterion" strength, which the soil improvement contractor is required to fulfil.

At present, there is still no well-accepted standard on how soil improvement specifications, including the "criterion" strength, should be derived from the design strength. For natural soils, guidelines such as those in the Eurocode 7 (BS EN 1997–1 2004) are available for determining a characteristic strength from the sample data. This approach cannot be used for improved soils which do not exist at the point of design. Instead, a common way is to specify a proportion or all of the core samples needed to equal or exceed a "criterion" strength in order for the soil improvement works to be considered adequate. In some cases, the minimum number of core samples required may also be specified.

An important question for the soil improvement engineer at the outset of chemical soil improvement works is "What is the binder type as well as binder and water contents required to fulfil the specifications of the chemical improvement works?" In cases where previous experience exists, the engineer can draw upon his experience. In cases where no previous experience exists, a workable recipe of binder and water contents is usually determined by laboratory moulding tests and field trials.

 DOI: 10.1201/9781003285885-5

5.2 LABORATORY TESTING OF MOULDED SAMPLES

Laboratory testing of moulded samples is often a precursor to the selection of trial mix since the laboratory strength is an indication of the "ideal" strength that can be achieved from a trial mix with high degree of uniformity and minimal air voids. The purpose of laboratory moulding tests is to ascertain a range or selection of prospective mix ratios which can potentially allow the specified strength or stiffness to be achieved. In instances where there is previous experience with the soil type in question, laboratory moulding tests may be obviated.

Further field trials are then conducted to determine a final mix ratio, which will be used in the actual works. The final mix ratio is not necessarily one which will give the specified strength. In fact, it is often a mix ratio that will lead to a strength which is higher than the specified strength. This is because the treated soil mixed in situ often exhibits significant spatial variation in strength and clients' specifications often require a majority of the samples or even all of the samples to equal or exceed the specified strength.

There are still no established guidelines on laboratory moulding procedures although some studies have been reported. An important issue in laboratory testing is the method of placing the cement–soil mix into the mould for curing. For instance, Grisolia et al. (2013) and Kitazume et al. (2015) conducted an international study to investigate the applicability of various methods of cement–soil placement, including placement without compaction, placement with tapping, rodding, static compaction (under different pressures) and dynamic compaction. The main findings from these studies are as follows:

1 The tapping method, which involved tapping the mould 50 times against the floor, gave the highest unconfined compressive strength for binder content below 40%. However, there is no specification on the impact force.
2 Samples without compaction or rodding show significantly lower strength and higher coefficient of variation, indicating poorer repeatability, than those with some form of densification or compaction treatment.
3 Rodding and tapping appear to give better repeatability than no compaction and dynamic compaction.
4 The repeatability of static compaction appears to differ, with the majority giving comparable coefficient of variation to rodding and tapping and a minority giving significantly higher coefficient of variation. This may indicate a dependence of the consistency on the compaction equipment.

Similar techniques were also used by other researchers. For example Xiao et al. (2014), Yao et al. (2018) and Ng et al. (2020) used a tapping approach, but instead of manual tapping, the samples were placed on a vibrating table which imparts vertical vibration. This may give slightly better consistency on the compaction effect than manual tapping. They also tested a few samples to ensure that the measurements are internally consistent. For instance, Ng et al. (2020) attributed sample-to-sample variability to air voids trapped in the sample. They measured the bulk density of the cured samples as a pre-test quality control indicator. If the measured bulk density deviates by more than ±1% of the target value, the specimen was discarded. In addition, the measured strength of the samples was monitored to ensure that they fall within a scatter range of 10%.

The other aspect of sample preparation, apart from moulding, is the mixing procedure. In general, the untreated soil is first remoulded to a standardised water content. Dry cement powder (Kitazume et al. 2015) or cement slurry (Xiao et al. 2014) are then mixed with the soil for about 10 minutes. This timing is in accordance with the guidelines of the Japanese Geotechnical Society (2009). Timings significantly longer than 10 minutes may overlap into the setting time of the cement. However, this limitation may not apply to non-cement binder, such lime or other pozzolanic binders, where setting time is not an issue.

The strength measured in laboratory tests is not necessarily a true reflection of the field strength since field processes differ markedly from laboratory procedures. In particular, most of the field mixing and grouting processes do not incorporate compaction. Furthermore, compressed air is used in some mixing processes to facilitate binder transport and penetration of mixing tools, as well as to shroud the cutting jet in double- and triple-tube jet grouting, in order to increase the cutting radius. In such cases, trapping of air voids is inevitable. Finally, the admixed soil usually has significant spatial variation in strength and specifications often require the criterion strength to be exceeded by all or a majority of the samples. Since laboratory-moulded samples are prepared under ideal conditions with minimal air voids and variability, laboratory-moulded sample strength is typically higher than the mean strength from field core samples, and much higher than the criterion strength, perhaps by a few times. Using data from 15 projects, Melancon and Shah (1973) reported that, in shallow soil–cement mixing for highway construction, significant discrepancies exist between laboratory and field strength. More than about 90% and 80% of laboratory-moulded and cured specimens achieved 7-day compressive strength of 225psi and 300psi, respectively. On the other hand, only about 35% and 15% of cored specimens achieved the same strength levels after 7 days of curing. Strength of field moulded specimens, which are collected by wet-grabbing freshly mixed soil in situ, shows closer agreement with cored sample strength, with about 47% and 26% achieving the same target strength. Hence, using laboratory-moulded and cured specimens alone may result in an optimistic assessment of the cored sample strength. Strength of laboratory-prepared specimens is better regarded as an upper limit of the achievable strength for that particular mix proportion. At present, there is still no widely accepted relationship between moulded sample strength, mean field strength and criterion strength.

5.3 ACCEPTANCE CRITERIA

Acceptance criteria are normally specified by the consultant or client, based on the level of performance of the treated soil that is required for the project and the importance of the project. Detailed criteria vary significantly between different regions of the world. However, most criteria will include specifications on method of quality assessment, density of coring per unit area of the treated soil domain or per unit length of linear treated domain, core recovery and minimum unconfined compressive strength. In some projects, wet-grab samples may also be required. Table 5.1

TABLE 5.1
Some typical acceptance criteria for production elements in deep mixing and jet grouting.

Source/project	Quality assessment/ sampling method	Density of sampling	Core recovery	Minimum strength (kPa)	Other conditions
		Deep soil mixing			
Puppala et al. (2008)	Wet-grab and vertically cored samples	Wet-grab samples: at least 1 sample per 750cu-yards of treated soil and at least 1 sample per day. Samples to be of sufficient size to provide 4 numbers of cylindrical specimens 3 inches diameter by 6 inches length for unconfined compressive test. 1 sample each to be tested after 7, 14 and 28 days curing. 4th sample to be tested after 56 days if 28 day strength fails to meet target. Cored samples: 15 random, continuous core samples with minimum diameter of 2 inches along the full element depth. 9 random specimens to be selected from core samples for unconfined compressive strength tests.		Minimum 14-day and 28-day unconfined compressive strength 100 psi (689.5kPa) and 1034.2kPa, respectively.	Minimum unit weight 105pcf (16.5kN/m^3). Maximum free vertical swell and linear shrinkage strain 1%.
Lemanza and Lesmana (2009) [Combined deep mixing and jet grouting]	Vertically cored samples	At per LTA (2010)	85%	Unconfined compressive strength ≥ 900kPa. Modulus E_{50} ≥ 90MPa	
EPA (2012)	Cored samples			Unconfined compressive strength ≥ 50psi (344.7kPa) for all treated cells or columns. Proportion of samples less than target value ≤ 10%. Permeability < 5 × 10^{-6} cm/sec for all treated cells or columns.	

(*Continued*)

TABLE 5.1 *(Continued)*

Some typical acceptance criteria for production elements in deep mixing and jet grouting.

Source/project	Quality assessment/ sampling method	Density of sampling	Core recovery	Minimum strength (kPa)	Other conditions
Bruce et al. (2013)	Vertical cored and wet grab samples	Cored samples: Full-depth core to be taken from 3% of treated elements or 79m^2 of treated area, whichever gives the larger number of cores. Maximum curing duration 28 days. Location: ¼ column diameter from column centre. Triple tube or equivalent coring method. Sample diameter ≥65mm, length ≥1m. Wet-grab samples: 1 bulk sample to be taken using bailer sampler or equivalent, every 2 production days or every 1500m^3 of treated soil, whichever gives the higher sampling frequency. 8 test specimens, each 76mm diameter by 152mm length, to be produced from each bulk sample. Testing to be conducted after 7 and 28 days of curing.	Dependent upon project requirements.		Laboratory permeability testing after 28 days curing (for production elements).
Daramalinggam and Yohannes (2019)	Cored samples	Minimum of 1 core per continuous block and/or 1 core per 1000m^3, taken between column centre and perimeter.	80% to 90%	Minimum unconfined compressive strength 0.6MPa to 1MPa (curing duration unknown, but often taken to be 28 days or longer).	Wet-grab sampling at depth for some projects.

LTA (2020)	Cored samples	4 boreholes for each 1,000m^3 of treated soil, midway between centre and perimeter of column. Core diameter should be at least 50mm. At least 1 sample to be taken from top and middle and bottom of each core for strength and stiffness testing, subject to minimum of 4 samples total.	85%. If TCR < 85%, two additional cores are to be taken at adjacent columns.	Depends on specific project requirements. If strength or stiffness targets are not achieved, two additional samples in the same core shall be tested for strength and stiffness.
Chew et al. (2020) for seepage cut-off wall	Cored samples	At least two cored locations per 100m section of the cut-off wall. Samples at 1.5m depth intervals to be taken along each selected DCM column. Unconfined compressive strength tests to be conducted for all cored samples. Permeability tests to be carried out on 5 samples per cored location.		Unconfined Compressive Strength (UCS) must lie between 200kPa and 450kPa. At least 70% of cored samples must have a UCS between 250kPa and 350kPa. Not more than 15% of the cored samples are to have a UCS between 200kPa and 250 kPa. Not more than 15% of cored samples are to have a UCS between 350kPa and 450kPa. Average permeability for 5 samples at each cored location < 10^{-9} m/s. Permeability of all samples must be ≤ 10^{-7} m/s.

(Continued)

TABLE 5.1 *(Continued)*
Some typical acceptance criteria for production elements in deep mixing and jet grouting.

Source/project	Quality assessment/ sampling method	Density of sampling	Core recovery	Minimum strength (kPa)	Other conditions
Some underground subway station and deep basement construction projects between 2015 and 2023	Cored samples	4 numbers of boreholes for each 1,000m^3. Location of the boreholes shall be where columns or augers overlap. Nominal core diameter ≥50 mm. At least one sample each to be taken from top, middle and bottom of each core for strength and stiffness testing. Unconfined compression tests and triaxial compression tests with strain measurement to be carried out to obtain strength and effective Young's modulus on the columns. Test results to be used to establish correlation between unconfined compressive strength (UCS) and Young's modulus.	Total core recovery (TCR) ≥ 85%. If TCR < 85%, 2 additional cores to the adjacent DSM columns to be carried. If either of the additional coring gives TCR < 85%, the DSM work is deemed to have failed.	(1) Minimum test result > 0.5 times stipulated value. (2) Mean of test results from at least 10 samples > 1.25 times stipulated value; OR 90% confident characteristic mean value determined from at least 10 samples > stipulated value. Examples of stipulated values: (a) Some subway station projects in Singapore: For 10–50 samples—Min. c_u = 400kPa (q_u = 800kPa), E_u = 140MPa. Mean c_u = 1030kPa, E_u = 350MPa. For more than 50 samples: Min. c_u = 400kPa (q_u = 800kPa), E_u = 140MPa. Mean c_u = 900kPa, E_u = 310MPa.	

		Jet grouting			
Force (2011)	Cored samples		85%	At least 90% of tested samples to achieve minimum 28-day unconfined compressive strength of 3.447MPa (500psi)	
Daramalinggam and Yohannes (2019)	Cored samples	Minimum of 2–4 cores over the length of column, every 1000m^3 of treated soil, at the overlaps.	80% to 90%	Minimum unconfined compressive strength 0.6MPa to 1MPa (curing duration unknown, but often taken to be 28 days or longer).	Wet-grab sampling at depth for some projects.
LTA (2020)	Cored samples and SPT	8 boreholes per 1000m^3 of treated soil, at overlaps. In each borehole, SPTs to be done at the top, middle and bottom of the treated zone. SPTs may be replaced by Cone Penetration Tests (CPTs) subject to acceptance; 4 cores through the depth of the jet grout pile per 1000m^3 of treated soil, at the overlaps. Core diameter should be at least 50mm. At least 1 sample to be taken from top and middle and bottom of each core for strength and stiffness testing.	85%. If TCR < 85%, two additional cores are to be taken at adjacent jet grout piles.	Depends on specific project requirements. If strength or stiffness targets are not achieved, two additional samples in the same core shall be tested for strength and stiffness.	

FIGURE 5.1 Wet-grab sampling box.
Courtesy of Keller Foundations (S E Asia) Pte Ltd.

shows some typical acceptance criteria used in chemical treatment of soil. The most commonly used method of quality assessment is by core sampling, typically by vertical cores. In some cases, wet-grabbing may also be used. In wet-grab sampling, the treated soil is sampled shortly after casting by inserting a sampling box into the treated ground (Figure 5.1). Upon retrieval, a trapdoor at the bottom of the box closes, thereby holding the soil within the box. The sample is then moulded, cured for the prescribed period and tested in the same way as a cored sample. Wet-grab sampling usually improves productivity since sampling can proceed in tandem with the treatment work. Double- or dual-tube sampling is sometimes also, albeit less often, used (Daramalinggam and Yohannes 2019) to sample soil from freshly treated ground. In this method, an outer cylindrical casing is driven into the ground, thus isolating the soil inside from that outside. A smaller inner casing is then inserted through the outer casing to collect the soil within the outer casing.

The density of cores is typically specified in terms of the number of cores per specified length/area/volume of the treated soil domain. If the configuration of the treated domain is largely linear, such as for cut-off or retaining walls, a linear measure is probably appropriate. For large area shallow treatment, an areal measure may be more appropriate. Finally, for massive or block-like treated domain, a volumetric measure may be used.

There may also be specifications regarding the locations of the cores. In particular, there is usually a preference for cores to be taken away from the centre of a deep-mixed or jet-grouted column, where there may be a concentration of grout and the treated soil may be strongest (see Section 3.6). This may, however, not be true for methods which allow grout to be introduced at the periphery of the cutting blades or a combined jet-grouting–deep-mixing approach (e.g. Raito 2007). Preferred regions for coring include midway between column centre and periphery, as well as overlap regions between adjacent columns. The overlap regions are most likely to be impacted if there are excessive errors in column positioning. Also specified are the sample size and number of samples to be tested from each core. At least three samples are typically tested from each core, with each sample having diameter of at least 50mm and a height-to-diameter ratio of 2.

The parameters which are commonly included in quality acceptance criteria are as follows:

1. **Total core recovery:** The total core recovery (TCR) is usually taken to be an indication of the continuity of the treated soil. Typically, minimum required TCR is between 80% and 90%.
2. **Unconfined compressive strength:** For foundation support and underground construction, the required minimum unconfined compressive strength usually ranges from 600kPa to 1MPa. For cut-off walls and other chemical stabilisation works, a lower minimum may be specified.
3. **Young's modulus:** For applications which are sensitive to stiffness, the acceptance criteria may also be Young's modulus. The modulus which is commonly specified is an "engineering" modulus measured by unconfined compression test, rather than the small-strain modulus.
4. **Permeability:** In special applications such as cut-off wall, the coefficient of permeability of the treated soil may not be permitted to exceed specified values (e.g. Chew et al. 2020).

Most of the acceptance criteria used to date are based on minimum strength or modulus. This usually implies that all tested samples must not have strength or modulus lower than the stipulated minimum. It precludes the use of a statistical framework, such as that used in BS EN 1997–1 (2004), where a characteristic strength is derived using statistical analysis. As will be discussed, the use of sample minima as a basis for assessment may lead to increased statistical fluctuation and error in quality assessment, especially if sample size is small.

5.4 PARAMETERS FOR SPATIAL VARIABILITY

5.4.1 Spatial Variation

Spatial variation usually arises due to two groups of factors, namely those relating to in situ variation of soil properties and those related to the quality of mixing. These have been discussed in Chapter 3. Spatial variability of deep-mixed and jet-grouted soil has been observed since the early days of these soil treatment methods. Kawasaki et al. (1984) measured direct shear strength of cored samples from boreholes closely spaced within clusters of four overlapping deep-mixed column. They reported direct shear strength ranging from less than 500kPa to about 2.5MPa. Chen et al. (2011) also reported significant variation in unconfined compressive strength of cored samples of cement-treated marine clay from two deep-mixing projects in Singapore. In the Marina Bay Financial Centre (MBFC) project, the unconfined compressive strength of approximately 480 cored samples ranged from about 800kPa to 5.2MPa. In the Nicoll Highway MRT Station project, the unconfined compressive strength of 128 cored samples ranged from about 1.1MPa to 5.1MPa. High strength variability has also been widely reported. For instance, Bruce et al. (2013) noted that "statistical analyses were performed on 7,873 unconfined compression strength tests from 14 datasets for 10 deep-mixing projects in the United States, and it was found that the coefficient of variation ranged from 0.34 to 0.79, with an average value of 0.56."

Similar variability has also been reported for jet grouting. Lee (1999) reported spatial variation in unconfined compressive strength in jet-grouted soil in high plasticity marine clay from three projects in Singapore, namely the Geylang River Widening project, Singapore River Reconstruction Contracts 1 & 2 and Singapore River Reconstruction Contract 3. In these three projects, the unconfined compressive strength ranges from about 1MPa to 3.2MPa, 400kPa to 4.1MPa and 300kPa to 3.8MPa, respectively. Other reports of spatial variation in jet-grouted soil include Mongiovì et al. (1991), Croce et al. (1994), Shibazaki et al. (1996), Croce and Flora (1998) and Nikbakhtan and Osanloo (2009).

5.4.2 Strength Distribution

The basic statistical parameters for quantifying spatial variation in soil properties, such as strength, are the mean (μ) and the coefficient of variation (CV). The latter is defined as the standard deviation (s_d) normalised by the mean, that is

$$CV = \frac{s_d}{\mu} \tag{5.1}$$

Although these two parameters are generally adequate for distributions which are normal or approximately so, spatial strength variation in cement-treated soils often do not follow a normal distribution. Honjo (1982) reported data from 17 deep-mixing projects in Japan. Twelve projects returned positive, that is left-skewed data, while four returned negative, that is right-skewed data. Only one project returned data with near-zero skewness, indicating a near-symmetric distribution. Lee's (1999) strength data from jet-grouted Singapore marine clay from the Geylang River Widening as well as Singapore River Reconstruction Contracts 1 to 3 also show a positive skew, which resembles a log-normal distribution. Filz and Navin's (2006) data from the Glen Road Interchange Ramps E & H also show a positive skew. Filz and Navin (2006) test-fitted 13 unconfined compressive strength data sets using the normal, triangular, uniform and log-normal distributions and noted that the log-normal distribution gave the best fit. Hence, in general, strength distribution from cement-treated soil is positively skewed and resembles the log-normal distribution. However, the minimum strength of treated soil samples is often higher than zero, This deviates from the log-normal distribution, which assumes a minimum strength of zero. Hence, there may be some fitting errors with a log-normal distribution at the lower end of the strength range.

The positive skew in strength distribution may be explained by the non-linear relationship between binder content and strength. Chen et al. (2016) developed a model for the statistical distribution of unconfined compressive strength, assuming that the in situ soil is uniform and that the grout (binder) content varies with the treated soil following a truncated normal distribution, together with a modified form of Lee et al.'s (2005) strength-binder relationship. Their results show that while the binder content may be normally distributed, the strength tends to be positively skewed, towards the lower strength end, which is akin to a log-normal distribution. This is due to the non-linear relationship between binder and water contents and strength. Hence, a significant factor affecting the strength distribution is likely to be the non-linear strength-binder relationship.

Liu et al. (2016) proposed that the β-distribution may permit a better description of range-bounded material properties since it enables the lower and upper limits of the ranges to be specified; this permits a non-zero minimum strength, as well as a maximum strength, to be fitted. The β-probability distribution function $f_x(x, \alpha, \beta)$ is given by

$$f_x(x,\alpha,\beta)=\begin{cases}\dfrac{(x-a)^{\alpha-1}(b-x)^{\beta-1}}{B(\alpha,\beta)(b-a)^{\alpha+\beta-1}}, & a\le x\le b\\ 0, & \text{otherwise}\end{cases} \tag{5.2}$$

in which a and b are the bounds, α and β shape parameters and $B(\alpha, \beta)$ is the β-function, given by

$$B(\alpha,\beta)=\frac{\Gamma(\alpha+\beta)}{\Gamma(\alpha)\Gamma(\beta)} \tag{5.3}$$

where $\Gamma(z)$ is the gamma function, given by $\Gamma(z)=\int_0^\infty t^{z-1}e^{-t}dt$.

Apart from range specifications, the shape of the β-distribution can also be adjusted via the variables α and β. Liu et al. (2015) also showed that, by using the β-distribution with $\alpha = 1.86$ and $\beta = 3.78$, a good fit can be obtained with the unconfined compressive strength distribution derived from field data from Phase 4 of the Marina Bay Financial Centre project in Singapore (Chen et al. 2011).

5.4.3 Mean and Coefficient of Variation

The mean unconfined compressive strength of treated soil varies substantially between projects since it depends upon the in situ soil type and water content as well as binder content, which depends, in turn, on the project requirement. The factors influencing the strength of the treated soil are discussed in Chapters 2 and 4. In general, for the same binder content, sandy and low plasticity soils tend to give higher treated strength than high plasticity and organic soils partly because their in situ water content tends to be lower and also because they require less water to facilitate mixing.

Table 5.2 shows data from some reported projects with reasonably large sample sizes, as well as some projects in which less commonly used mixing and sampling methods were employed. This includes a case by Chen et al. (2011) in which a jet mechanical mixing method was employed. Honjo's (1982) coefficient of variation (CV) values are narrowly ranged between 0.30 and 0.39, although duration of curing varies between 30 and 100 days. On the other hand, Filz and Navin's (2006) data from 13 projects show CV ranging more widely, from 0.34 to 0.79, with an average of 0.57. After accounting for strength variation arising from variation in curation duration and water-to-cement ratio of the slurry, Filz and Navin (2006) reported a much lower average CV of 0.39. Nonetheless, the adjusted CV still ranges from approximately 0.17 to 0.67, which remains much larger than Honjo's (1982). In some of Filz and Navin's (2006) projects, there are also significant differences between the unadjusted and adjusted CV. For instance, for the owner's wet-grab samples from the Jackson Lake Dam, the unadjusted CV is about 0.79. However, the adjusted CV is only about 0.26.

The data from Lee (1999), Chen et al. (2011), Liu et al. (2017) and Daramalinggam and Yohannes (2019) reflect the experience in Southeast Asia, primarily Malaysia and Singapore. With the exception of Chen et al.'s (2011) data from the Nicoll High MRT Station, the unadjusted CV for all the other projects range from 0.28 to 0.52, which is smaller than that reported by Filz and Navin (2006), but larger than that of Honjo's (1982). Chen et al.'s (2011) Nicoll Highway MRT Station data were obtained from soil improved by the jet mechanical mixing (JMM) or RAS-JET system (Raito 2007). As Figure 5.2 shows, this combines mechanical mixing with high pressure grout jets from nozzles located at the ends of the mixing blade. It should also be noted that the binder content used, at least in the deep-mixing part, ranges from 400 to 600kg/m^3. This is much higher than the typical cement content of 180kg/m^3 used in projects in Singapore. It is also much higher than the binder contents reported by Honjo (1982). This might have an indirect effect on the CV.

Table 5.2 indicates that CV values can vary widely over different regions. Babasaki et al. (1997) proposed that CV can range from 0.22 to 0.27. Honjo's (1982) data, which reflect Japanese conditions, indicate CV ranging from 0.30 to 0.39. In Singapore and Malaysia, CV values of 0.42 to 0.5 are commonly encountered. There appears to be no significant differences in CV deep mixing and jet grouting. On the other hand, Filz and Navin's (2006) data show wide-ranging CV, which appears to be roughly consistent with Larsson's (2003) observation that CV can range from 0.2 to 0.6. Notwithstanding this, with sufficient experience, it may be possible to define a range of CV variation for a particular soil type or condition, which is reflective of the workmanship.

TABLE 5.2

Some unconfined compressive strength parameters from field data.

Reference/project	Number of samples	Method of sampling	Mean (kPa)	Unadjusted coefficient of variation (CV)
Honjo (1982)/Hiroshima Port	236	Core	3840	0.35
Honjo (1982)/Yokohama Port Block no. 1 (clay)	346	Core	4780	0.36
Honjo (1982)/Yokohama Port Block no. 1 (sand Depth 22m–25m)	138	Core	4760	0.39
Honjo (1982)/Yokohama Port Block no. 4 (clay)	209	Core	4630	0.30
Honjo (1982)/Yokohama Port Block no. 4 (sand)	129	Core	5210	0.32
Honjo (1982)/Tokyo Bay L-6	114	Core	2090	0.38
Filz and Navin (2006)/Oakland Airport	184	Core	3537	0.4
Filz and Navin (2006)/Glen Road Interchange G & F	184	Core	3261	0.34
Filz and Navin (2006)/Glen Road Interchange H & E	634	Core	2840	0.42
Filz and Navin (2006)/Blue Circle/Kinder Morgan Cement Silos	487	Wet grab	4702	0.74
Filz and Navin (2006)/Jackson Lake Dam (owner's samples)	355	Core	4640	0.67
Filz and Navin (2006)/Jackson Lake Dam (owner's samples)	1018	Wet grab	2296	0.79
Filz and Navin (2006)/Jackson Lake Dam (contractor's samples)	1710	Wet grab	2785	0.65
Filz and Navin (2006)/I-95 Interchange single auger	473	Core	2840	0.67
Filz and Navin (2006)/I-95 Interchange multiple auger	2199	Core	3420	0.66
Liu et al. (2017)/Marina Bay Financial Centre	232	Core	1700	0.42
Liu et al. (2017)/Marina One	1145	Core	2100	0.44
Chen et al. (2011)/Nicoll Highway MRT Station (Jet mechanical mixing in marine clay layer)[1]	51	Core	3880	0.24[1]
Lee (1999)/Geylang River Widening (jet grouting in marine clay)		Core	1843	0.30
Lee (1999)/Singapore River C1 & C2 (jet grouting in marine clay)		Core	1500	0.52
Lee (1999)/Singapore River C3 (jet grouting in marine clay)	130	Core	1770	0.45
Daramalinggam and Yohannes (2019)/Site A Malaysia	51	Core	3680	0.43
Daramalinggam and Yohannes (2019)/Site E-P1 Malaysia	113	Core	3890	0.46
Daramalinggam and Yohannes (2019)/Site E-P2 Malaysia	86	Core and double tube	3300	0.28
Daramalinggam and Yohannes (2019)/Site F Malaysia (jet grouting)	104	Backflow sampling[2]	4880	0.42
Daramalinggam and Yohannes (2019)/Site G Malaysia	82	Double tube	1700	0.49

1 A deep-mixing method termed Jet Mechanical Mixing (JMM) was used. In this method, the central part of the column was improved by mechanical mixing and the peripheral part by high pressure jet grouting (Osborne and Ng 2008). The cement content was also much higher than typical deep-mixing projects in Singapore. These factors led to relatively high mean strength and a negatively skewed strength distribution that is quite atypical.

2 Backflow sampling refers to the collection of samples from the jet-grouting backflow comprising a mixture of soil, grout and water.

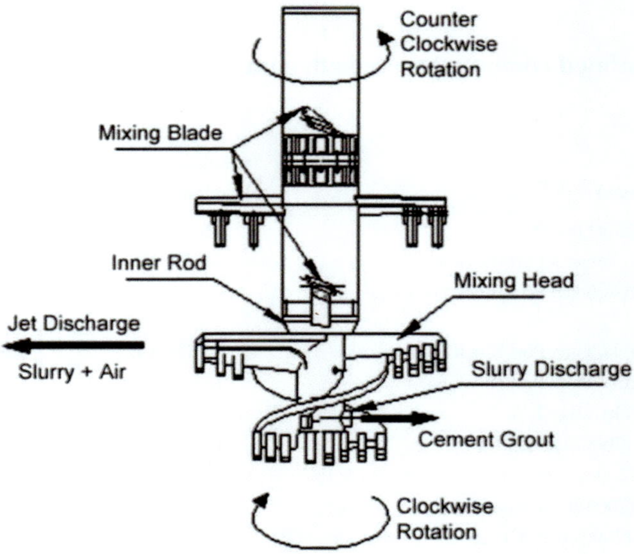

FIGURE 5.2 Schematic of a Jet Mechanical Mixing (JMM) mixing tool
Source: **Chen et al. 2011.**

5.4.4 Scale of Fluctuation

Another important descriptor of a random field is the autocorrelation structure, which characterises the "persistence" in the random field. This refers to a tendency in which points close together will have similar properties whereas points located far apart tend to have quite different properties (Fenton and Griffiths 2010). Autocorrelation structures of random fields can be expressed by an autocorrelation function and autocorrelation length. In geotechnical engineering, the autocorrelation structure is often assumed to be stationary; that is it does not change within a site. This allows the autocorrelation function to be characterised by a single parameter, namely the autocorrelation length or scale of fluctuation (SoF).

The SoF is the distance within which a soil property, such as strength, of a geological material is statistically similar or correlated. The correlation between the values of a random variable q at coordinates x and $x+\tau$, where τ (≥ 0) is the separation distance, is expressed by the autocorrelation function $\rho(\tau)$, defined by

$$\rho(\tau) = \frac{E\left(\left(q_{x+\tau} - \mu(q)\right)\left(q_x - \mu(q)\right)\right)}{\sigma^2(q)} \tag{5.4}$$

where E is the expected or mean value operator and $\mu(q)$ and $\sigma^2(q)$ represent the mean and variance of q. The function $E\left(\left(q_{x+\tau} - \mu(q)\right)\left(q_x - \mu(q)\right)\right)$ is also known as the auto-covariance function $K(\tau)$. The value of the autocorrelation function $\rho(\tau)$ must lie in the range [−1, 1]. $\rho(\tau) = 1$ implies perfect correlation, whereas $\rho(\tau) = -1$ implies perfect anti- or inverse correlation. $\rho(\tau) = 1$ when $\tau = 0$ and generally decreases as τ increases. If $q_{x+\tau} = q_x$, then the numerator becomes $E\left(\left(q_x - \mu(q)\right)^2\right) = \sigma^2(q)$ and $\rho(\tau) = 1$. Numerous functions have been proposed to fit $\rho(\tau)$. Examples of commonly used autocorrelations functions include the following (Vanmarcke 1983).

5.4.4.1 Exponential Function

One of the most commonly used autocorrelation functions is the exponential function, which is given by

$$\rho(\tau) = \exp\left(-\frac{2\tau}{\theta}\right) \tag{5.5}$$

where θ is the SoF (or autocorrelation length) and is greater than or equal to 0. When $\tau = \theta$, then $\rho(\tau) = \exp(-2) \sim 0.135$. Hence in the exponential function, the SoF is the distance at which the autocorrelation coefficient decays to approximately 0.135.

Honjo (1982) used a slightly different form of the exponential function, given by

$$\rho(\tau) = \exp\left(-\frac{\tau}{\Theta}\right) \tag{5.6}$$

where $\Theta = \theta/2$ is termed an autocorrelation distance. In the following discussion, Equation (5.5) will be used instead, as it is more commonly adopted.

5.4.4.2 Squared Exponential Function

Another commonly used autocorrelation function is the squared exponential function, also known as Gaussian function, and is given by

$$\rho(\tau) = \exp\left(-\pi\left(\frac{\tau}{\theta}\right)^2\right) \tag{5.7}$$

When $\tau = \theta$, then $\rho(\tau) = \exp(-\pi) \sim 0.043$. Hence in the squared exponential function, the SoF is the distance at which the autocorrelation coefficient decays to approximately 0.043.

5.4.4.3 Triangular Function

The triangular autocorrelation function is given by

$$\rho(\tau) = \begin{cases} 1 - \dfrac{\tau}{\theta} & \text{for } 0 \le \tau \le \theta \\ 0 & \text{for } \tau \ge \theta \end{cases} \tag{5.8}$$

When $\tau = \theta$, then $\rho(\tau) = 0$. Hence, in the triangular function, the SoF is the distance at which the autocorrelation coefficient decays to 0. Other autocorrelation functions which have been reported include cosine exponential, second-order Markov and spherical functions. The cosine exponential is an exponentially decaying cosine function and is suitable for special situations in which the autocorrelation trend has a harmonic component superimposed onto the trend of decay.

5.4.5 Determination of Scale of Fluctuation

Honjo (1982) computed depth-averaged autocorrelation coefficients from several deep-mixing projects in Japan using the following relationship:

$$\rho(\tau) = \frac{1}{D}\int_D \frac{\left(q_u(z) - \overline{q_u}\right)\left(q_u(z+\tau) - \overline{q_u}\right)}{\sigma^2(q_u)} dz \tag{5.9}$$

where $q_u(z)$ is the unconfined compressive strength at depth z, $\bar{q}_u$ the average unconfined compressive strength and D the total depth of borehole. If q_u is measured at n regular and discrete depth intervals Δz over the depth of the borehole, then Equation (5.9) can be expressed in discrete form as

$$\rho(\tau) = \frac{1}{D}\sum_{i=1}^{n} \frac{\left(q_{ui}(z) - \overline{q_u}\right)\left(q_{ui}(z+\tau) - \overline{q_u}\right)}{\sigma^2(q_{ui})} \Delta z \tag{5.10}$$

where $q_{ui}(z)$ and $q_{ui}(z+\tau)$ refer to q_u of the ith sample and that of the sample that is offset at distance τ from it, respectively. Detrending may also be applied to the data prior to estimating the SoF. This is particularly relevant for vertical SoF of parameters, such as water content, undrained shear strength or cone resistance, which is known to decrease or increase with depth. To date, there is no hard-and-fast rule on data detrending and much depends upon the judgement of the analyst. The trend may not be readily distinguishable from the SoF and presence of the trend can significantly affect not only the SoF but also the CV.

Figure 5.3 shows fitting attempts to the "Group A" data from Honjo (1982), who reported that these Group A data are from relatively uniform ground conditions, with mixing parameters that reflected standard industry practices. The Hiroshima Port data are reasonably fitted using a triangular autocorrelation function with SoF of 5m. However, the Tokyo Bay L-11 and L-13 data appear to follow an exponential decay. The squared exponential function is unable to reflect the rapid decrease in autocorrelation coefficient at small offset τ. Honjo (1982) bracketed the data using the exponential function with SoF ranging from 4m to 8m.

Liu et al. (2019) evaluated horizontal and vertical SoFs for the Marina Bay Financial Centre (MBFC) and Marina One projects. For the vertical SoF, Liu et al. (2019) used Honjo's (1982) depth-averaging approach. Table 5.3 shows the site, soil

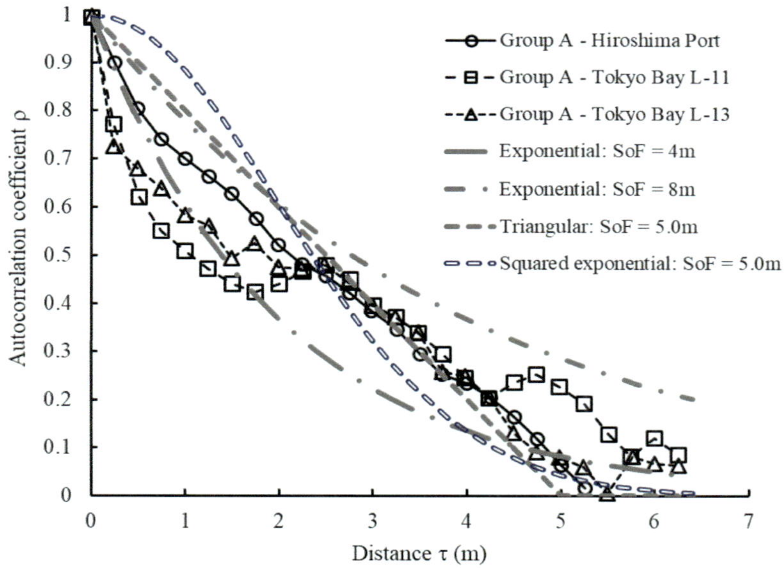

FIGURE 5.3 Vertical autocorrelation data from some construction projects in Japan and fitted autocorrelation function.

Refitted using data from Honjo (1982).

and mixing parameters for these two projects. Since the separation distance τ varied significantly from one borehole to another, Liu et al. (2019) subdivided the vertical separation distances into bands, for example less than 3m, 3m to 4m, 6m to 7m and so on.

Liu et al. (2019) used a similar approach for the horizontal autocorrelation coefficient. Horizontal separation distances are first into bands, for example less than 20m, 20m to 30m, 30m to 40m and so on. Autocorrelation coefficients from all boreholes are then sorted into these bands of separation distances. The average autocorrelation coefficient is then computed for each band of separation distances based on the available samples. The representative separation distance for each band is taken to be the average separation distance of the samples within that band.

Figures 5.4(a) and (b) show the fitted autocorrelation functions for in situ water content of the soil pre-treatment, using Liu et al.'s (2019) data. As Figure 5.4a shows, the vertical water content autocorrelation data, albeit limited, are bounded by single exponential functions with SoF of 4m to 8m. This can be attributed to the close proximity of the two projects sites, which are separated by about 100m. On the other hand, the horizontal autocorrelation data show greater discrepancy and Liu et al. (2019) fitted the MBFC and Marina One data using SoFs of 100m and 200m, respectively, which would indicate significantly different ground characteristics. However, re-examination of the data suggests another possible explanation. At separation distances smaller about 80m, the autocorrelation coefficient of both sites appears to indicate an SoF of about 100m, together with a significant amount of fluctuation. For separation distances larger than

TABLE 5.3
Site, soil and deep-mixing parameters of Marina Bay Financial Centre (MBFC) and Marina One projects.

Parameter	Site: Marina Bay Financial Centre (MBFC)	Site: Marina One
Site dimensions and size	Approximately 190m × 181m (~34,000m²)	Approximately 169m × 166m (~28,000m²)
Depth of improved soil (m)	14–22	15–25
Depth of marine clay (m)	8–40 (Upper marine clay)	12–24 (Upper marine clay) 18–48 (Lower marine clay)
In situ marine clay properties		
Mean water content (within improved region) (%)	69.1	Upper marine clay: 56 Lower marine clay: 31
Mean plastic limit (%)	30	Upper marine clay: 29 Lower marine clay: 24
Mean liquid limit (%)	80	Upper marine clay: 69 Lower marine clay: 53
Deep-mixing parameters		
Type of cement	Ordinary Portland Cement (CEM I)	Portland Blast Furnace Cement (CEM III)
Specific gravity of cement	3.2	3
Unit weight of cement slurry (g/cm³)	1.56	1.5
Binder content per m³ of soil: (kg/m³)	280	180
Water–cement ratio of grout	0.9	1
Number of columns installed per jack-in-withdrawal sequence	4	2
Number of blades per shaft	4	4
Blade rotation number (blade-revs/m)	279.4	325.6
Diameter of deep-mixed column (m)	1	1.3
Nominal area covered by each jack-in-withdrawal sequence of deep-mixing equipment (m²)	3.61	2.48
Cement grout discharge rate (litres/min)	480	255
Discharge time per metre length of column (min)	2.54	2.33
Blade rotational speed (RPM)	27.5	27.5
Required unconfined compressive strength (MPa)	0.8	0.8
Number of core samples	232	1145

Data source: Liu et al. 2019.

about 80m, there is better agreement between the autocorrelation coefficient of the two sites, and a SoF of between 210m and 330m is implied.

Figure 5.5 shows the autocorrelation coefficients of the two sites for separation distances below 100m, together with the sample sizes (in square brackets) from which the autocorrelation is estimated. For separation distances below 50m, the sample sizes were generally smaller than those for larger separation distances. A minimum

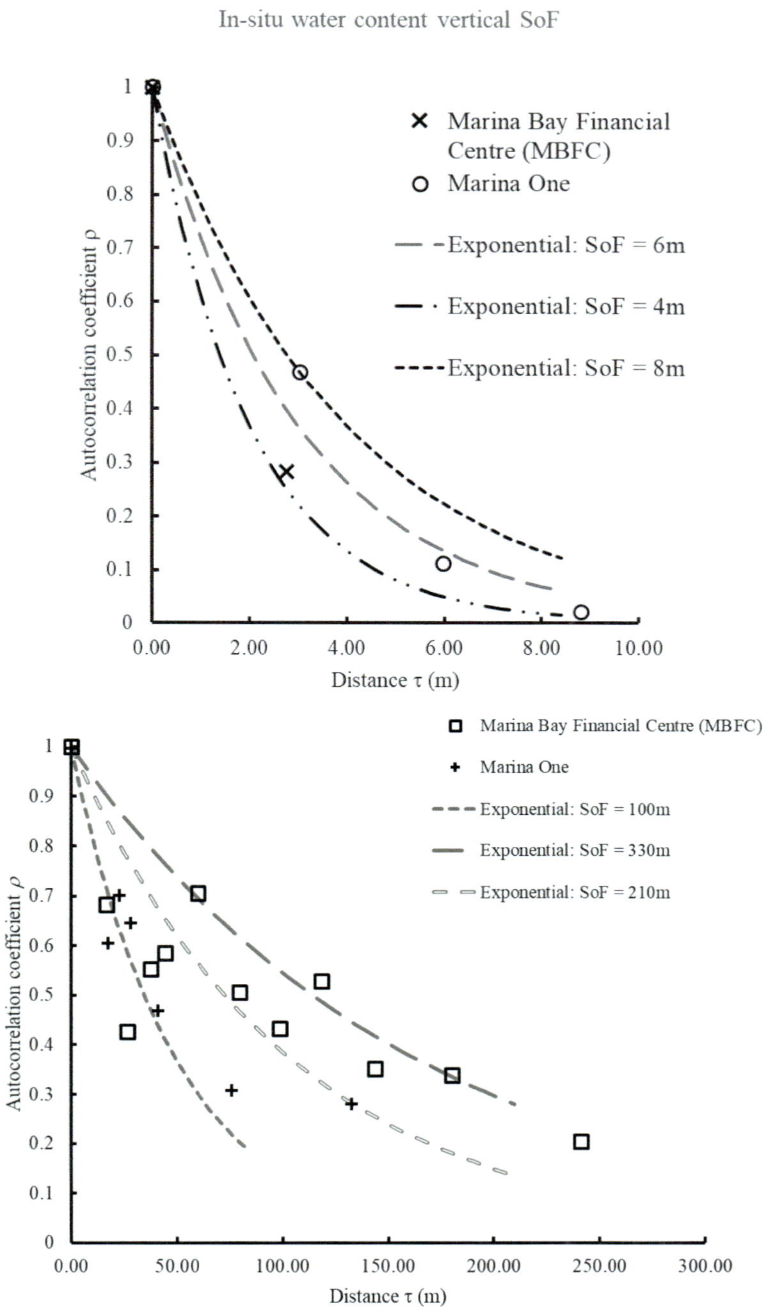

FIGURE 5.4 (a) Vertical and (b) horizontal autocorrelation functions for in situ water content in ground improvement works at the Marina Bay Financial Centre (MBFC) and Marina One projects. **Replotted from Liu et al.'s (2019) data.**

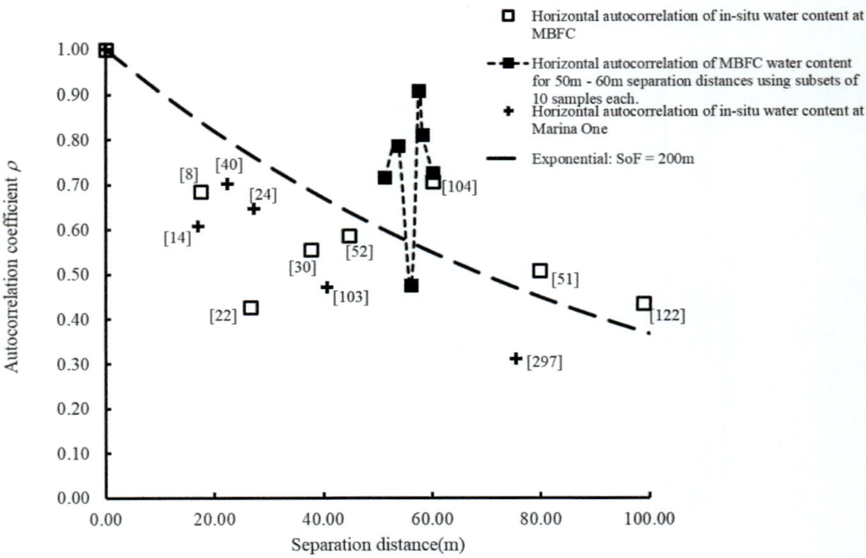

FIGURE 5.5 Horizontal autocorrelation coefficients for in situ water content in the MBFC and Marina One projects for separation distances less than 100m. Numbers in square brackets [] denote the sample sizes used in the estimation of the respective autocorrelation coefficients.

Refitted from Liu et al.'s (2019) data.

sample size is generally required in correlational estimates. For instance, Fraenkel et al. (2023) proposed that a sample size of at least 50 is required for correlational studies. Huitema and McKean (1991) also noted that sample sizes below 50 can lead to biased estimates in autocorrelation estimates. Figure 5.5 also illustrates the effect of using smaller sample sizes. The data for MBFC includes a set of 60 samples with separation distances between 50m and 61m. To assess the effect of smaller sample sizes, these samples were sorted in order of increasing separation distances and then sub-divided into sub-sets of 10 samples each. Each sub-set therefore represents a smaller range of separation distance and this is reflected by the closer spacing of the resulting autocorrelation points. The data sub-sets show a significant increase in the level of fluctuation between successive autocorrelation coefficients, with values ranging from approximately 0.9 to 0.5. Hence, a possible effect of insufficient sample size is an increase in the level of fluctuation in the autocorrelation estimates.

In many practical situations, the sample size may not reach the minimum requirement of ~50, and some fluctuation may be inevitable. Large sample sizes may be obtained by combining samples from different parts of a site, as is done in MFC and Marina One projects. This is equivalent to taking average autocorrelation estimates across the entire site, the implicit assumption being that the autocorrelation characteristics are invariant over the entire site. In such cases, local knowledge of the site and workmanship may be helpful in deciding how samples ought to be combined.

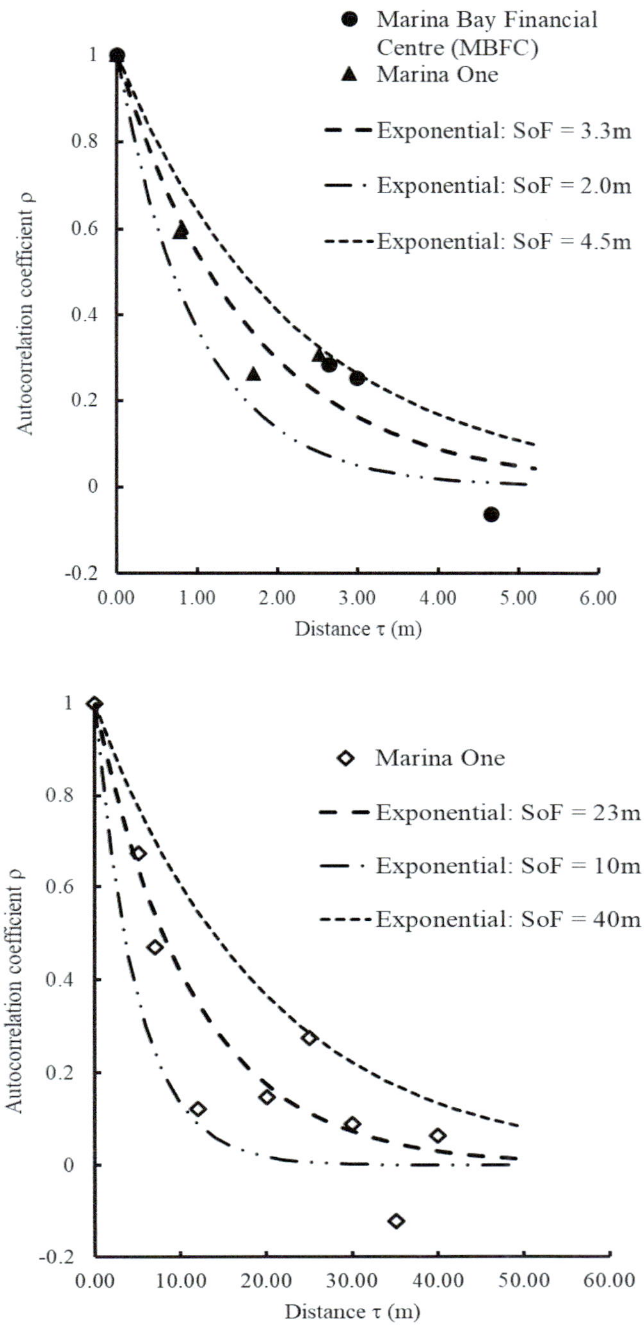

FIGURE 5.6 (a) Vertical and (b) horizontal autocorrelation functions for treated soil strength at the Marina Bay Financial Centre (MBFC) and Marina One projects.

Refitted using Liu et al.'s (2019) data.

As Figure 5.6 shows, the unconfined compressive strength shows similar attributes as the in situ water content, with the vertical SoF being much shorter than the horizontal SoF. Both SoFs are much smaller than the corresponding SoF for in situ water content. This is due to superposition of the random spatial variation arising from imperfect mixing. SoF associated with deep mixing is typically much shorter than that of in situ water content. This is because each treated soil column is mixed independently of the others and therefore spatial variation between columns should not correlated. The superposition of this mixing-induced short-range spatial variation onto the long-range water content variation has the overall effect of reducing the autocorrelation coefficient for the same lag distance, which, in turn, reduces the SoF.

5.5 EFFECT OF CURING DURATION ON STATISTICAL CHARACTERISTICS OF MEASURED STRENGTH

Filz and Navin (2006) conducted regression analyses and noted that the CV is significantly influenced by the water–cement ratio and curing time up to the point of testing, the latter hereafter termed "curing duration". The effect of water–cement ratio (and binder content) on strength has been discussed in Chapter 4. In many chemical improvement projects, the same water–cement ratio and binder content are used for entire sites or large parts of a site in question, especially where the soil to be treated is of the same kind. Where different soil types are to be treated or large variations in water content are expected within the site or project, it may be advisable to assess the spatial variation of these soil types separately since different soils may have different in situ water content and miscibility characteristics and require different binder contents for the same gain in strength.

In many projects, there are no clear guidelines on the curing time-to-testing. Filz and Navin (2006) did not provide any details on how their regression analysis was conducted. Liu et al. (2017) presented a simple method for normalising the strength with respect to the curing duration, using data from the MBFC and Marina One projects. In both projects, the curing duration varies from 20 days to 180 days. To prevent masking of the curing duration effect by the random scatter due to spatial variation, Liu et al. (2017) used the average strength for each curing period to allow the hardening trend to be better manifested. The hardening trend was then fitted using Equation (4.1-5a) in Table 4.1:

$$q_u = q_\infty \left[\frac{(\alpha t / q_\infty)^r}{1 + (\alpha t / q_\infty)^r} \right] \left\{ \frac{\exp[m(s/c)]}{(w/c)^n} \right\} \tag{5.11}$$

This relationship assumes that the unconfined compressive strength increases with the curing duration t according to a generalised hyperbolic relationship. If the average soil–cement and water–cement ratios can be assumed to be constant for the different curing periods, Equation (5.11) can be re-written as follows:

$$\frac{1}{q_u} = \frac{1}{q_1}\left[1 + \frac{1}{\left(\alpha_1 t / q_1\right)^r}\right] \tag{5.12}$$

where $q_1 = q_\infty \left\{\exp\left[m(s/c)\right]/(w/c)^n\right\}$ and $\alpha_1 = \alpha\left\{\exp\left[m(s/c)\right]/(w/c)^n\right\}$ are both constants. If this relationship holds, then a plot of $\frac{1}{q_u}$ versus $\frac{1}{t^r}$ should be a straight line with a gradient of $\frac{1}{\alpha^r q_1^{1-r}}$. The value of r was first adjusted until points fell on a reasonable straight line. The intercept and gradient then allow q_1 and α_1 to be inferred.

Figures 5.7a and b show the raw strength data, average strength and fitted strength-curing-duration curves for the MBFC and Marina One projects. The raw strength data show considerable scatter. However, the average strength for different curing durations shows much less scatter and permits the trend to be better discerned. For MBFC, the general uptrend in average strength is very slight, indicating that the strength has largely stabilised after a curing period of about 50 days. This is consistent with the completion of the hydration reaction of the ordinary Portland cement (Chew et al. 2004). For Marina One, the uptrend is more evident. This can be attributed to the shorter minimum curing period and the use of Portland blast-furnace cement. Portland blast furnace cement CEM III (EN 197–1 2012) is a blend of Ordinary Portland cement and blast furnace slag. Depending upon the sub-classification (A, B, C) the percentage of blast furnace slag can range from 35% to as high as 95%. The inclusion of blast furnace slag provides substantial amounts of lime, silica and alumina, which facilitated significant long-term pozzolanic reaction and higher long-term strength, while the reduced proportion of Portland cement leads to lower early strength. As Table 5.3 shows, the binder content used in Marina One is significantly lower than that in MBFC. Despite this, the long-term average strength in Marina One is higher.

Liu et al. (2017) noted that normalising the strength with respect to curing duration has a discernible influence on the spread of the strength data, although the CV was not significantly affected in the two projects. In many projects, curing duration is not stipulated as a condition for core testing. Where the binder used is known to produce significant long-term pozzolanic reaction, some specifications on the curing duration may help to remove uncertainty from the interpretation of the core sample data. It may also be unnecessary to follow the 28-day rule, which is widely used in structural concrete. This is because the lead time between ground treatment and excavation or loading often exceeds 28 days. In such cases, a longer curing duration will allow the effect of pozzolanic reaction to be better assessed.

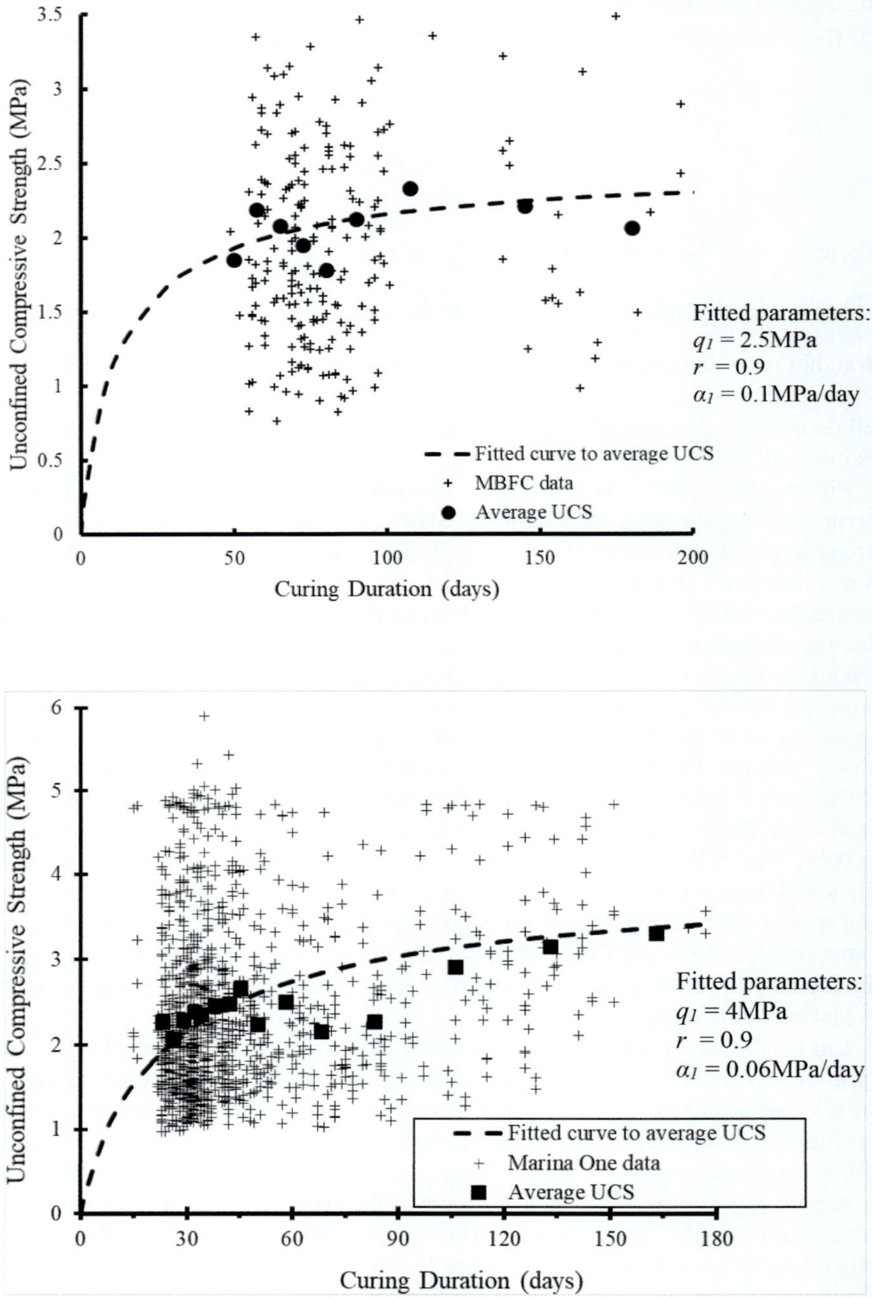

FIGURE 5.7 Raw data, average and fitted curve of unconfined compressive with curing duration for (a) MBFC and (b) Marina One projects.

Data source for average strength data: Liu et al. (2017).

5.6 SCALE OF FLUCTUATION FOR NATURAL AND CEMENT-TREATED SOILS

Table 5.4 shows some reported SoFs for natural and cement-treated soils. For natural soils, SoFs are most widely reported for water content. In general, the vertical SoF of water content of natural soils ranges from less than 1m to about 13m. This is much smaller than the horizontal SoF, which is typically larger than 50m. Pieczyńska-Kozłowska (2015) reported vertical SoF for the cone penetrometer resistance of layered silty and clay soils, ranging from 0.07m to 0.29m, which is even smaller than that for water content. This is not surprising since cone resistance is a strength parameter and can be influenced by factors such as over-consolidation ratio as well as soil type and structure, apart from water content. Pieczyńska-Kozłowska (2015) also noted that the vertical SoF is heavily influenced by the method of detrending prior to fitting of the correlation model. The largest SoF was obtained by using linear detrending, that is by subtracting a linear trend of water content decreasing with depth. The smallest SoF was obtained by using a combination of linear and quadratic trends to match the CPT data. The latter removes much of the longer range SoF, leaving the short-range fluctuations. This suggests that, in many practical scenarios, it may be difficult to distinguish SoF from a complex trend.

For chemically improved soils, Honjo (1982) reported vertical SoF ranging from 0.8m to 8m in sandy soils and soft clays from deep-mixing projects in Japan. Honjo (1982) did not mention if detrending was applied to the data prior to SoF determination. Liu et al. (2019) estimated a vertical strength SoF of approximately 3.3m based on data from MBFC and Marina One. Liu et al. (2019) also reported vertical binder mass fraction SoF of 1m to 3.3m, based on centrifuge model deep-mixing tests on normally consolidated kaolin clay. Liu et al. (2019) did not apply any detrending to their data. Both their strength and binder content SoF lie within the range reported by Honjo (1982). This suggests that, for wet deep mixing, vertical strength (and binder mass fraction) SoF may range typically from 0.8m to 8m. In contrast, Al-Naqshabandy et al. (2012) and Bergman et al. (2013) reported much smaller vertical strength SoF ranging from approximately 0.08m to 0.77m for dry lime/cement mixing. This may be due to the reduced flowability of the material phases in dry mixing processes, which limits their mixing distances. Al-Naqshabandy et al. (2012) and Bergman et al. (2013) did not mention if detrending was applied to their data. In view of this, it is reasonable to surmise that no detrending was applied in all the earlier-cited vertical SoF analyses for chemically improved soils.

Horizontal SoF of chemically improved soils shows much greater variance between studies. Larsson et al. (2005) reported short-ranged SoF in both radial and circumferential directions within individual dry lime/cement mixed columns. Liu et al. (2019) also reported similar short-ranged SoF from centrifuge model studies of individual wet-mixed columns. On the other hand, Navin (2005) and Liu et al. (2019) reported much longer-ranged SoF of between 24m and 37m, based on field-data analyses of wet deep-mixing projects. Finally, intermediate-ranged SoFs from 1.8m to 3.6m were also reported by Larsson and Nilsson (2009), Al-Naqshabandy et al. (2012) and Bergman et al. (2013).

TABLE 5.4

Some reported SoFs of natural and improved soils.

Parameter	Reference	Soil type or treatment method	Scale of fluctuation (m)	
			Vertical	Horizontal
In situ water content	Diaz-Padilla (1974)	Soft organic silty clay	2.4–3.7	–
	Alonso and Krizek (1975)	Chicago drift soil	<0.7	–
		London clay	<2.5	–
		Mexico City clay	<1.6	–
		Soft silty loam	<5.0	–
	Western and Grayson (1998)	Tarrawarra catchment in south-eastern Australia	–	70–100 (wet winter period) 100–120 (dry summer period)
	Phoon and Kulhawy (1999)	Clay, loam	1.6–12.7	170 (average for clay)
	Akbas and Kulhawy (2010)	Ankara clay	2.5–5.5	–
	Grote et al. (2010)	Sandy loam and clay loam	–	60–100
	Liu et al. (2019)	Singapore marine clay (MBFC and Marina One sites)	6	100–200
	Valdez-Llamas et al. (2003) (Gulf of Mexico)		Superficial soft clay 0.8–2.0 Deeper layers 21	Superficial soft clay 22 Deeper layers 1000
Cone penetrometer resistance	Pieczyńska-Kozłowska (2015)	Silty and clayey soil (multi-layered) from Świebodzice	0.07–0.29[1]	–
	Cami et al. (2020)		0.26–0.52[2]	–

Binder mass fraction in chemically admixed soil	Liu et al. (2019). Centrifuge model tests on deep mixing	Kaolin clay	1.0–3.3m (average ~2.2m)	Radial direction: 0.12–0.28D, where D = diameter of deep-mixed column. (Average 0.17D). Circumferential direction: 67°–133° (average 89°). D = 2.4m in Liu et al.'s study.
Strength of improved soil	Honjo (1982)	Variety of soft soils from deep-mixing projects in Japan (both sand and clay)	0.8–8.0	–
	Larsson et al. (2005)	Dry lime/cement mixing	–	Radial direction: <0.22D. Circumferential direction: < 0.32m (angular SoF < 59°).
	Navin (2005)	Wet deep mixing	–	24–37
	Larsson and Nilsson (2009)	Dry lime/cement mixing	–	1.8–3.6
	Al-Naqshabandy et al. (2012)	Dry lime/cement mixing	0.2–0.7	2.0–3.0
	Bergman et al. (2013)	Dry lime/cement mixing	0.08–0.77	< 3.5
	Liu et al. (2019)	Singapore marine clay (MBFC and Marina One sites)	3.3	25

1 Range obtained primarily from using different methods of detrending.

2 Range obtained using linear detrending and different correlation models.

5.7 SOME FACTORS AFFECTING THE MEASURED SPATIAL VARIATION IN TREATED SOIL STRENGTH

Chen et al. (2016) developed a statistical model for the unconfined compressive strength of deep-mixed columns, which assumes that spatial variation in the strength of the admixed is caused solely by imperfect mixing. The outline of this model is as follows:

(a) The starting point of Chen et al.'s (2016) is centrifuge data on spatial variation of binder mass fraction within a wet deep-mixed column (Figure 3.6), which is defined as the ratio of the mass of binder, such as cement slurry (i.e. cement powder and water used to form the cement slurry), to the total mass of the admixed soil. It can be readily shown that the soil–cement ratio x and water–cement ratio y of the admixed soil can be related to the water–cement ratio of the cement slurry a and the binder mass fraction b by the following relationships:

$$x = \frac{1}{A}\left(\frac{1}{b} - 1\right) \tag{5.13a}$$

$$\text{and} \quad y = wx + a \tag{5.13b}$$

where w is the in situ water content of the soil and

$$A = \frac{1+w}{1+a} \tag{5.13c}$$

The strength is deduced using Equation (4.1–5a), that is

$$q_u = q_0 \frac{\exp(mx)}{y^n} = q_0 \frac{\exp(mx)}{(wx+a)^n} \tag{5.14}$$

$$\text{in which } q_0 = q_\infty \left\{ 1 - \frac{1}{1 + \left(\frac{\alpha t}{q_\infty} \right)^r} \right\} \tag{5.15}$$

Chen et al. (2016) noted that the exponential term in the numerator of Equation (5.14) will cause the strength to tend to infinity at very low cement content, which is counterintuitive. Hence, they replace the exponential term *exp*(*mx*) by the first three terms of the Taylor series expansion. Substituting in Equations (5.13a) and (5.13b) leads to

$$r = \frac{q_u}{q_0} = \frac{1 + \frac{m}{A}\left(\frac{1}{b} - 1\right) + \left\{\frac{m}{A}\left(\frac{1}{b} - 1\right)\right\}^2}{\left\{\frac{w}{A}\left(\frac{1}{b} - 1\right) + a\right\}^n} \tag{5.16}$$

where r is the normalised unconfined compressive strength. Chen et al. (2016) assumed that the water–cement ratio of the slurry, a, is a constant as the slurry is pre-mixed in mixing tanks and quality of mixing can be more assured. The random variable is the binder mass fraction b, which is affected by in situ mixing with the soil.

(b) The probability density function of normalised unconfined compressive strength, $f_R(r)$, can be related to the probability density function of binder mass fraction, $f_B(b)$, by

$$f_R(r) = \left|\frac{db}{dr}\right| f_B(b) \tag{5.17}$$

(c) The binder fraction b is assumed to follow a truncated normal distribution in the interval [0, 2μ] and coefficient of variation, following Figure 3.6. This allows the probability distribution function of the unconfined compressive strength to be evaluated.

As Figure 5.8 shows, Chen et al.'s (2016) single random parameter model was able to replicate the positive skewness in the MBFC strength distribution as well as many strength distributions of improved soils (see sub-section 5.4.2). Chen et al. (2016) attributed this positive skewness to the non-linear relationship between strength

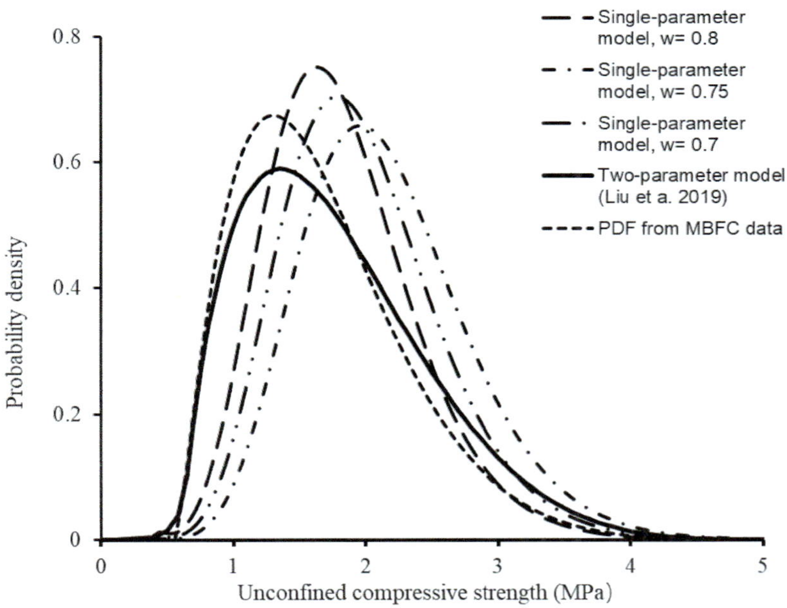

FIGURE 5.8 Comparison of Chen et al.'s (2016) single-parameter model (with different in situ water content w) and Liu et al.'s (2019) two-parameter model with field data from MBFC. **After Liu et al. (2019).**

and binder concentration. There are also differences between Chen et al.'s (2016) calculated strength distribution and that deduced from MBFC data. The computed probability distribution function is peakier and narrower than the measured probability density function, despite using different in situ water contents, indicating that the calculated probability distribution function has a smaller CV than the measured function. Chen et al. (2016) postulated that the differences between calculated and observed strength distribution functions might have been caused by two possible factors. The first is the variability in the curing duration. As the curing duration increases, the strength distribution curve shifts to the right, indicating a general increase in strength. One may therefore surmise that, if the curing duration varies from sample to sample, the strength distribution will be more dispersed; that is there will be an increase in CV of the strength distribution. This increase in CV is spurious and can be minimised by minimising the variation in curing duration. For Ordinary Portland Cement, Liu et al. (2017) noted that strength increase will largely cease after about 60 days. Hence, using a curing duration longer than 60 days is also likely help minimise curing duration effects.

The second factor is the spatial variation in the in situ water content. They noted that, in the MBFC site, the water content is noticeably higher in the upper one-third of the improved depth, compared to the top and bottom boundaries. This can be attributed to the on-going consolidation of the marine clay due to previous land reclamation work at this location. Liu et al. (2019) extended Chen et al.'s (2016) single-parameter model to a two-parameter model, which considers natural variation in soil properties, in addition to spatial variation in binder distribution due to imperfect mixing. The horizontal SoF of variation due to imperfect mixing should be less than one column diameter in the radial direction and less than one column circumference in the circumferential direction, since mixing is done column by column and each column is mixed independently of the others. Liu et al. (2019) postulated that one possible source for the long range reported SoFs is natural variability in soil properties, especially water content. Using data from MBFC and Marina One sites, Liu et al. (2019) showed that, as the water content of the natural marine clay decreases from about 75% at 10m depth to about 45% at 25m depth, the 28-day strength of the treated marine clay, averaged over depth intervals of 2m, increases from about 1.6MPa to about 2.4MPa. Using Equation (5.16), Liu et al. (2019) was able to predict the increase in strength and obtained reasonably good agreement with measured 2-m depth averaged strength. Furthermore, as Figure 5.8 shows, Liu et al.'s (2019) two-parameter model was able to generate a probability distribution function which is closer to that of the MBFC data than Chen et al.'s (2016) single-parameter model.

To investigate the effect of horizontal water content variation on the SoF of the unconfined compressive strength of the treated soil, Liu et al. (2019) used the following numerical procedure:

1. Generate realisations of the random fields of in situ water content and the binder mass fraction using Monte-Carlo simulations.
2. Randomly sample the spot in situ water content and the binder mass fraction from the generated random fields.

3 Calculate the unconfined compressive strength from the sampled in situ water content and the binder mass fraction using Equation (5.16).
4 Repeat steps (a) to (c) for 10,000 realisations to ensure the curve of correlation coefficient with lag distance, that is the correlation curve, is stable.

Liu et al.'s (2019) results showed that the short-range spatial variation arising from imperfect mixing interacts with the long-range spatial variation in in situ water content in a complex manner. Both component CVs contribute towards the overall CV of the treated soil, but in a weighted root-sum-square fashion rather than direct superposition (Figure 5.9). The SoF of the treated soil strength will be intermediate between that of the short-range mixing-induced SoF and long-range in situ water content SoF (Figure 5.10). Hence, the effect of short-range mixing-induced SoF is to cause the overall SoF to be smaller than in situ water content SoF, as previously discussed.

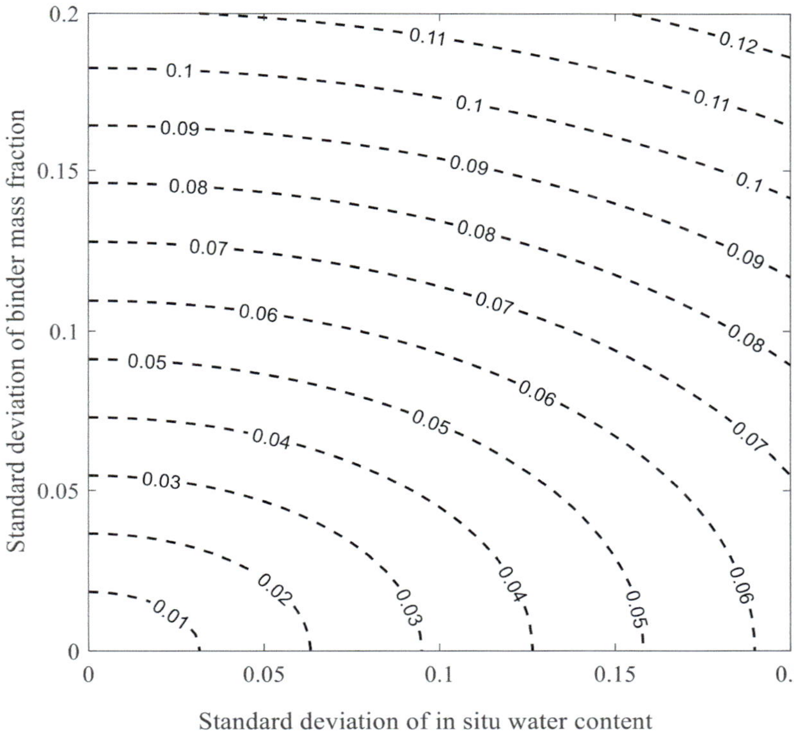

FIGURE 5.9 Relationship between standard deviation for strength ratio *r*, in situ content and binder mass fraction, based on the mean values of in situ water content and binder fraction in the MBFC project. Values of contours are strength ratio *r* [see Equation (5.16)].

After Liu et al. (2019).

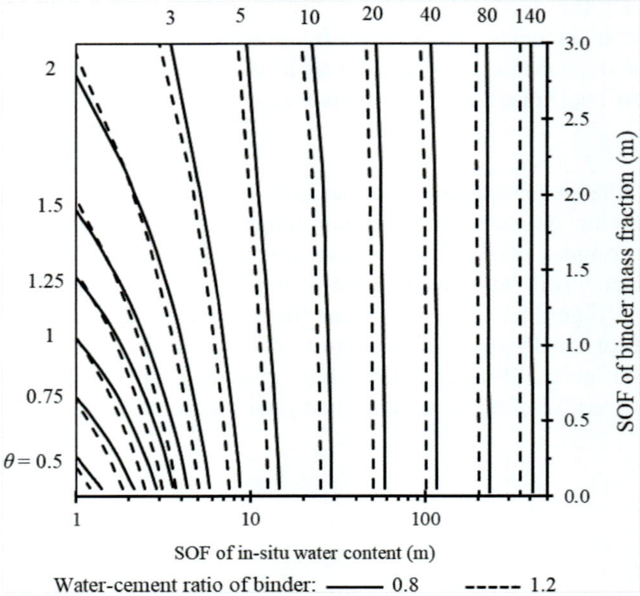

FIGURE 5.10 Relationship between scales of fluctuation for strength of treated soil θ, binder mass fraction b and in situ water content w, for water–cement ratio of binder $a = 0.8$ and 1.2. **After Liu et al. (2019).**

5.8 PRACTICAL IMPLICATIONS FOR QUALITY CONTROL

The findings of this study have implications for the quality control of deep-mixing operations. Firstly, the CV of the treated soil strength will be higher than the CV due to imperfect mixing, owing to spatial variation in in situ water content. Liu et al. (2019) developed a relationship between the CV of the strength of the treated soil with the CVs of binder mass fraction and in situ water content. However, the equation is relatively complex and may not be readily applicable in practice. In such cases, one alternative is to conduct part of the core samples at short intra-column distances or at least distances that are much smaller than the SoF of in situ water content variation. This ensures that the water content within the sampled area remains relatively constant. Another option may be to use the strength CV of the treated soil as an estimate. This is viable if the water content does not vary significantly over the site, for example if the site is small or the water content is relatively constant. In addition, if several ground improvement contractors are operating on the same site, the CV of the treated soil strength may be useful as a comparative indicator of their workmanship. A smaller CV will imply more uniform mixing and thus better workmanship. Hence, evaluation and monitoring of the CV should be encouraged as an important component of the quality assessment framework.

Liu et al.'s (2019) results also highlighted the co-existence of SoF at intra- or inter-column levels. In large cement treatment projects, the dimension of the treatment area is likely to be of a similar order or greater than the SoF of in situ water content. Moreover, the SoF of in situ water content is typically much greater than that of variation in binder concentration. The presence of two very different SoFs has implications for core sampling, quality control and simulation of cement-treated soil. The long-range SoF may be important for overall performance of a cement-treated soil mass, whereas the short-range SoF has a significant influence on the strength and stiffness of individual column.

In core sampling for quality control of cement deep-mixing works, the spacing between boreholes is typically much greater than a column diameter. For instance, in Marina One, the typical borehole spacing is about 10m. At such spacings, the mean and CV of the spatial distribution may still be reflected, but the short-range intra-column SoF, which is more reflective of the mixing quality, may not be. One way to ameliorate this is to sink some of the boreholes at close spacing of less than a column diameter, if intra-column variation is to be captured. Conversely, in simulating random fields for cement-treated ground, using an SoF that is much greater than a column diameter would imply that the strength within each column is quite uniform, which may not be realistic.

Whereas the SoF may require care in its assessment and interpretation, the CV is a relatively simple parameter to evaluate. Furthermore, notwithstanding the effect of in situ water content variation on the CV of the treated soil, the CV remains a useful tool for comparing workmanship on site. This is especially so when different soil improvement contractors work on the same site or in close proximity, and effect due to in situ water content variation can be largely discounted.

An important issue which is often encountered in ground improvement projects is the minimum number of core samples to be measured. As Table 5.1 shows, the minimum required number of core samples are typically specified in relation to the volume, area or length of the ground domain to be treated. Large treatment projects may involve thousands of core samples and there is often no issue with the minimum sample number. For small projects, the number of samples may be much smaller. In some cases, the number of samples may be in the region of 20 to 30. In such cases, a question may arise as to whether the number of samples is sufficiently representative of the spatial variation on the site.

This issue was examined by Liu et al. (2017) using data from MBFC and Marina One projects, using a statistical bootstrapping approach (Bourdeau and Amundaray 2005). This approach assumes that the data sets from the two projects are sufficiently large to be representative of the population characteristics. By repeatedly and randomly selecting sub-sets (consisting of a prescribed smaller number of samples) from the overall data set, one can examine how pre-selected strength parameters, such as median, mean, 5th percentile strength and so on, fluctuate with different randomly selected samples. These strength parameters can be considered to be representative strength parameters for each sub-set, since they characterise the statistical attributes of that particular sub-set. For instance, by randomly selecting data sub-sets of, say, 20 samples each, one can examine how the median, mean or 5th percentile strength fluctuates with different sub-sets. The results can be expressed as a CV of the corresponding representative strength parameters. Furthermore, by progressively increasing the sample size, that is number of data points in the sub-sets, one can also examine how these strength parameters changes with sample size. As a normal occurrence, the strength parameter should generally stabilise to a constant value as sample size increases. This is because when the sample size is equal to the population size, there should be no further fluctuations in any of representative strength parameters; that is the CV of all representative strength parameters should be zero.

Liu et al.'s (2017) pre-selected representative strength parameters included the sample minima, 5th, 10th, 50th percentile and the (mean—$k \times$ standard deviation) strength, where k is a multiplier ranging from 0 to 1.3. With the exception of the sample minima, the CV of all the other representative strength parameters decrease as the sub-sample size increases, indicating that the corresponding representative strength parameter stabilises progressively as the sub-sample becomes larger. The decrease in the representative strength CVs was rapid at small sub-set size, but moderates when the sub-set size exceeds about 300. In contrast, the CV of the sample minima initially increases with sub-set size before decreasing. This anomalous behaviour occurs because the sample minima is an outlier value; increasing the sub-set size can initially increase the chance of encountering even more extreme outlier values, which, in turn, increases the CV of sample minima. In terms of quality assessment, this implies that using the sample minima as a strength criterion potentially increases the likelihood of encountering extreme outlier (low) strength values as the number of samples increases. This can increase the likelihood of disputes unnecessarily and discourage soil improvement contractors from collecting more samples, which run counter to the aim of quality assessment. In view of this, the sample minima is not recommended for use as a strength criterion in quality assessment. It is better to set a low percentile strength, such as 5th percentile or (mean—$k \times$ standard deviation) strength, as a representative strength parameter in quality assessment.

5.9 GEOPHYSICAL METHODS

To date, the most commonly used quality assessment and control method for chemically admixed soils remains strength testing of wet and dry core samples. However, core sampling is not merely laborious; it only provides statistical information on the spatial variation of the improved soil. It does not provide any information on the specific locations and sizes of weak zones and voids. In contrast, geophysical measurements can potentially allow specific locations and stiffnesses of strong and weak zones within the treated domain to be determined.

Tatsuoka (2010) reported the shear modulus of cement-treated soils deduced from field-measured shear wave velocities. However, apart from providing a check on the overall stiffness of the treated ground, no quality assessment was conducted using seismic measurements.

There are relatively few studies relating to geophysical measurements for quality monitoring of deep-mixed and jet-grouted soils. Madhyannapu et al. (2010) used spectral analysis of surface waves to determine the average shear wave velocity profiles with depth of two sites comprising 3-m long lime–cement–soil columns installed in a non-contiguous layout. They showed that the average shear wave velocity of the treated areas is higher than that in the untreated areas but lower than that obtained by laboratory tests, as the field-measured shear wave velocity represents an average value of the composite ground comprising treated soil columns and inter-column untreated regions. Average downhole P-wave velocity in soil-mixed columns was also significantly higher than that in untreated locations. However, stiffness variation within the treated domains was not investigated.

Lindh and Lemenkova (2023) used uphole seismic field measurements to evaluate the depth profiles of P- and S-wave along a lime- and cement-stabilised soil column. A hole was drilled through the centreline of a stabilised soil column. A seismic source was lowered into the hole and excitation was triggered at prescribed depths. Accelerometers were installed on the soil surface level of the column to measure acceleration in the x- and y-horizontal as well as vertical directions. This allows the wave velocity of SH- and P-waves to be captured for different source depths, thus allowing the profile of wave velocity with depth to be inferred.

Mackens et al. (2015) also used cross-borehole P-wave geotomography to verify the diameter of jet-grouted columns and assess the effects of curing time on soil stiffness. The measurements indicate that geotomography can be used to define the outlines of the stiffer treated soil columns within softer untreated soil domain. However, none of the aforementioned works dealt with the detection and characterisation of random weak zones.

Bearce (2015) conducted a proof-of-concept study on the feasibility of using P-waves to detect zones of different stiffnesses in a jet-grout column, using 1-g model tests. The model test setup consisted of a single cement-admixed soil column with a diameter of approximately 20cm to 25cm. The crosshole sonic logging source and receiver systems consisted of two 5cm-diameter water-filled PVC pipes, each containing a hydrophone transceiver with 45kHz centre frequency. The depth of the transceivers can be adjusted by cable reels. Travel time was determined from first arrivals and data inversion performed using a commercial seismic processing software. A simultaneous iterative reconstruction technique (SIRT) algorithm was used to derive the ray paths. Bearce's (2015) results indicate the presence of strong and weak layers as well as open pockets of weak material. However, the results could not be verified since no controlled specimens with known shapes and sizes of weak zones were tested. There is also a requirement that the water pipes containing the transceivers must be in contact with or embedded inside the improved soil column. Finally, the 45kHz frequency is far higher than the frequency range which is typically used in seismic surveys. In a typical cement treatment scenario, the column diameter is typically 2m or less. If the ratio of column diameter to wavelength is to be maintained, wave frequency ranging from 4.5kHz to 5.6kHz would be required for prototype treated soil columns; this is much higher than the frequency range typically used in seismic surveys.

More recently, Wong et al. (2025) also used 1-g model tests to study the feasibility of detecting a circular weak zone of known size within a cement-treated soil block using SH-waves and travel time tomography using the SIRT algorithm. The model set-up consists of a cement-treated soil block with an 80mm-diameter cylindrical weak zone cast in-place (Figure 5.11), the latter simulating a "plane-strain" weak zone. Two vertical arrays of bender elements were mounted on opposing faces of the soil block as transceivers. Single pulse sinusoidal waves with frequency up to 35kHz were triggered from the source benders. At 35kHz, the wavelength of the waves in

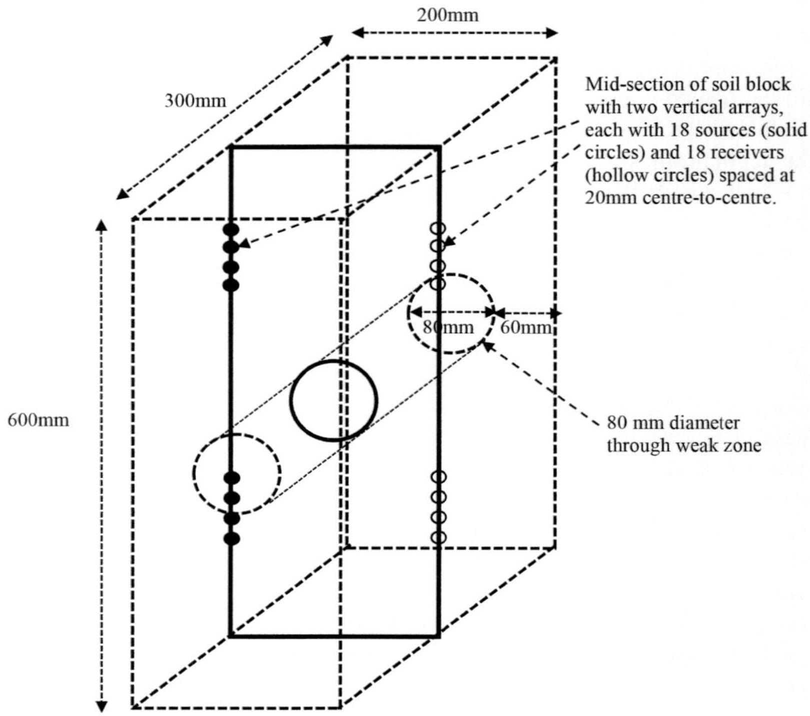

FIGURE 5.11 Wong et al.'s (2025) 1g-model setup, showing outline of soil block (dashed lines) and mid-section (in full lines) with transceiver arrangement.

the cement-treated soil (not weak zone) is roughly one-fifth of the diameter of the weak zone. The benders were laterally offset from the edges of the weak zone by a distance of at least 60mm. This simulates a scenario in which seismic sources and receivers were installed at a close proximity to the weak zone.

Wong et al.'s (2025) results show that while the location, shape and size of the weak zone can be reasonably ascertained using travel time tomography with first arrivals, the stiffness of the material within the weak zone may be significantly over-estimated, especially in cases where the stiffness of the weak zone is substantially lower than that of the surrounding soil. This is because the first arrivals are dominated by waves that diffract around weak zone. This masks the arrival of the transmitted waves through the weak zone, which contain information on the weak zone stiffness. The diffracted waves travel a slightly longer path around the weak zone, but travel at the speed of the stronger surrounding soil. These diffracted waves give a spuriously high wave speed, leading to over-estimation of the stiffness.

Wong et al. (2025) showed that, by extracting the 35kHz Morlet wavelet components from the source and received signals using wavelet transform and estimating the travel time of this wavelet component through the time lags of the wavelet in the source and received signals, stiffness can be reliably estimated, down to a much lower stiffness ratio, that is ratio of stiffness of weak zone to that of the surrounding soil. This is because the 35kHz Morlet wavelet has much less low frequency content that the raw received signal. It is therefore much less susceptible to diffraction around the weak zone. For a prototype weak zone diameter of, say, 1m, maintaining the ratio of weak zone diameter to wavelength would require a wave frequency of 2.8kHz. This is still much higher than frequency ranges typically used seismic surveys. Hence, measuring the stiffness of metre-sized weak zones in cement-treated soil may still pose significant challenges with existing seismic setups.

Studies relating to use of georesistivity techniques for quality assessment of chemically admixed soils are even scarcer. Bearce (2015) reported a model study on the use of ring electrodes surrounding a cement-treated soil column for measurement of column diameter. However, the study did not extend to the measurement of hardened stiffness or strength of the treated soil. Ng et al. (2022) proposed a framework for relating the electrical conductivity of cement-treated soil to its strength. However, to date, this framework has not been applied to field studies.

This discussion shows that geophysical methods are not widely used and probably have not reached the required level of maturity. A critical limitation lies in the current equipment. In order to detect metre-sized weak or strong inclusions, the wavelength of should be no longer than about one-fifth the inclusion size (Ávila-Carrera and Sánchez-Sesma's 2006). For the typical range of wave speed encountered in cement-treated soil, this may require frequency ranging from 2kHz to 4kHz. On the other hand, conventional geophones used in seismic surveys typically operate on frequencies no higher than about 40Hz, although higher frequency sensors are available, which can reach the kHz range. Another issue is that high frequency components are preferentially attenuated by the ground than low frequency components. Hence, a source which can radiate sufficient energy is required. In short, what is required are high frequency signal sources of sufficient strength and receivers.

REFERENCES

Akbas, S. O. and Kulhawy, F. H. (2010). Characterization and estimation of geotechnical variability in Ankara clay: A case history. *Geotechnical and Geological Engineering*, 28(5): 619–631.

Al-Naqshabandy, M. S., Bergman, N. and Larsson, S. (2012). Strength variability in lime-cement columns based on cone penetration test data. *Proceedings of the Institution of Civil Engineers—Ground Improvement*, 165(1): 15–26. https://doi.org/10.1680/grim.2012.165.1.15.

Alonso, E. E. and Krizek, R. J. (1975). Stochastic formulation of soil properties. *Proc. 2nd Int. Conf. Applications of Stat. and Probability Soil & Struct. Engrg.*, Aachen, Germany: 9–32.

Ávila-Carrera, R. and Sánchez-Sesma, F. J. (2006). Scattering and diffraction of elastic P- and S- waves by a spherical obstacle: A review of the classical solution. *Geofísica Internacional*, 45(1): 3–21. https://doi.org/10.22201/igeof.00167169p.2006.45.1.189.

Babasaki, R., Terashi, M., Suzuki, T., Maekawa, A., Kawamura, M. and Fukazawa, E. (1997). Japanese geotechnical society technical committee report—factors influencing the strength of improved soil. *Proc. IS Tokyo '96/2nd Int. Conf. Ground Improvement Geosystems—Grouting and Deep Mixing*—, Balkema, Rotterdom: 913–918.

Bearce, R. G. (2015). *Geometry assessment and strength/stiffness monitoring of lime and cement modified soils via characterization of curing-induced property changes estimated from seismic wave propagation techniques and electrical resistivity.* PhD dissertation, Colorado School of Mines.

Bergman, N., Al-Naqshabandy, M. S. and Larsson, S. (2013). Variability of strength and deformation properties in lime cement columns evaluated from CPT and KPS measurements. *Georisk*, 7(1): 21–36.

Bourdeau, P. L. and Amundaray, J. I. (2005). Non-parametric simulation of geotechnical variability. *Géotechnique*, 55(2): 95–108. http://dx.doi.org/10.1680/geot.2005.55.2.95.

Bruce, M. E. C., Berg, R. R., Collin, J. G., Filz, G. M., Terashi, M. and Yang, D. S. (2013). *Federal highway administration design manual: Deep mixing for embankment and foundation support.* McLean, VA, USA: Department of Transportation Federal Highway Administration.

BS EN 1997–1 (2004). *Eurocode 7: Geotechnical design—Part 1: General rules.* European Committee for Standardization.

Cami, B., Javankhoshdel, S., Phoon, K. K. and Ching, J. (2020). Scale of fluctuation for spatially varying soils: Estimation methods and values. *ASCE-ASME Journal of Risk and Uncertainty in Engineering Systems, Part A: Civil Engineering*, 6(4): 03120002.

Chen, J., Lee, F. H. and Ng, C. C. (2011). Statistical analysis for strength variation of deep mixing columns in Singapore. In *GeoFrontiers 2011: Advances in geotechnical engineering, Dallas, TX* (Eds. J. Han and D. E. Alzamora), GSP 211, Geo-Institute of American Society of Civil Engineers, Reston, VA, USA: 576–584.

Chen, J., Liu, Y. and Lee, F. H. (2016). A statistical model for the unconfined compressive strength of deep-mixed columns. *Géotechnique*, 66(5): 351–365.

Chew, S. H., Kamruzzaman, A. H. M. and Lee, F. H. (2004). Physicochemical and engineering behavior of cement treated clays. *Journal of Geotechnical and Geoenvironmental Engineering*, 130(7): 696–706.

Chew, S. H., Yim, H. M. A., Koh, J. W., Chua, K. E. and Gng, Z. X. (2020). QA/QC plan and evaluation of field strength and permeability value for deep cement mixing. *International Journal of GEOMATE*, 19(74): 52–58.

Conner, J. R. (1994). Chemical stabilization of contaminated soils. In *Hazardous waste site soil remediation—theory and application of innovative technologies* (Eds. D. J. Wilson and A. N. Clarke), CRC Press, Boca Raton.

Croce, P. and Flora, A. (1998). Effects of jet grouting in pyroclastic soils. *Rivista Italiana di Geotecnica*, Patron editore, (2): 7–24. ISSN 0557-1405.

Croce, P., Gajo, A., Mongiovì, L. and Zaninetti, A. (1994). Una verifica sperimentale degli effetti della gettiniezione. *Rivista Italiana di Geotecnica*, 28(2): 91–101 (in Italian).

Daramalinggam, J. and Yohannes, M. M. (2019). Specifications, sampling methods and correlations of engineering properties for Deep Soil Mixing and Jet Grouting in Malaysia. *Proc. Int. Conf. Case Histories Soil Prop.*, Singapore.

Diaz-Padilla, J. (1974). *Settlement of structures on shallow foundations: A probabilistic analysis.* PhD thesis, Massachusetts Institute of Technology, Cambridge, MA, USA.

EN 197–1 (2012). *Cement—Part 1: Composition, specifications and conformity criteria for common cements.* European Committee for Standardization.

EPA (2012). *U.S. EPA removal action Walton and Lonsbury site Attleboro, Massachusetts: Performance specification for soil mixing for ground improvement.* Rev. 2, 7 March 2012, https://semspub.epa.gov/work/01/524003.pdf.

Fenton, G. A. and Griffiths, D. V. (2010). Reliability-based geotechnical engineering. In *GeoFlorida 2010: Advances in analysis, modelling and design* (Eds. D. O. Fratta and A. J. Puppala), ASCE, Geotechnical Special Publications, FL.

Filz, G. M. and Navin, M. P. (2006). Stability of column-supported embankments. *Final Contract Report, VTRC 06-CR13*, Virginia Transportation Research Council, Charlottesville, VA.

Force, J. G. T. (2011). Port of Galveston pier 16 & 18 jet grouting specification, American Society of Civil Engineers, Reston, VA, USA.

Fraenkel, J. R., Wallen, N. E. and Hyun, H. H. (2023). *How to design and evaluate research in education*, 11th Edition, McGraw Hill.

Grisolia, M., Leder, E. and Marzano, I. P. (2013). Standardization of the molding procedures for stabilized soil specimens as used for QC/QA in deep mixing application. *Proc. 18th Int. Conf. Soil Mech. Geotech. Eng.*, Paris: 2841–2484.

Grote, K., Anger, C., Kelly, B., Hubbard, S. and Rubin, Y. (2010). Characterization of soil water content variability and soil texture using GPR groundwave techniques. *Journal of Environmental and Engineering Geophysics*, 15(3): 93–110.

Honjo, Y. (1982). A probabilistic approach to evaluate shear strength of heterogeneous stabilized ground by deep mixing method. *Soils and Foundations*, 22(1): 23–38.

Huitema, B. E. and McKean, J. W. (1991). Autocorrelation estimation and inference with small samples. *Psychological Bulletin*, 110(2): 291–304.

Japanese Geotechnical Society (2009). *Practice for making and curing stabilized soil specimens without compaction*, JGS T 0821-2009 (in Japanese).

Kawasaki, T., Saitoh, S., Suzuki, Y. and Babasaki, R. (1984). Deep mixing method using cement slurry as hardening agent. *Seminar on Soil Improvement and Construction Techniques in Soft Ground*, Singapore: 17–38.

Kitazume, M., Grisolia, M., Leder, E., Marzano, I. P., Correia, A. A. C., Oliveira, P. J. V., Ahnberg, H. and Andersson, M. (2015). Applicability of molding procedures in laboratory mix tests for quality control and assurance of the deep mixing method. *Soils and Foundations*, 55(4): 761–777.

Larsson, S. (2003). *Mixing processes for ground treatment by deep mixing*. PhD thesis, Royal Institute of Technology, Stockholm, Sweden, ISSN 1650-9501.

Larsson, S. and Nilsson, A. (2009). Horizontal strength variability in lime-cement columns—a field study. *Proc. Int. Sym. Deep Mixing & Admixture Stab.* (Ed. M. Kitazume), Port and Airport Research Institute (PARI), Okinawa, Japan: 629–634.

Larsson, S., Stille, H. and Olsson, L. (2005). On horizontal variability in lime-cement columns in deep mixing. *Géotechnique*, 55(1): 33–44. https://doi.org/10.1680/geot.2005.55.1.33.

Lee, F. H., Lee, Y., Chew, S. H. and Yong, K. Y. (2005). Strength and modulus of marine clay-cement mixes. *Journal of Geotechnical and Geoenvironmental Engineering*, 131(2): 178–186.

Lee, Y. (1999). *A framework for the design of jet grout piles in Singapore marine clay*. MEng dissertation, National University.

Lemanza, W. and Lesmana, A. (2009). Deep soil improvement technique using combined deep mixing and jet grouting method. *Proc. 17th Int. Conf. Soil Mech. Geot. Engrg.* (Eds. Mamza, Shahien and El-Mossallamy), IOS Press: 2439–2442.

Lindh, P. and Lemenkova, P. (2023). Ultrasonic P- and S-wave reflection and CPT soundings for measuring shear strength in soil stabilized by deep lime/cement columns in Stockholm Norvik Port. *Archives of Acoustics*, 48(3): 325–346. https://doi.org/10.24425/aoa.2023.145243.

Liu, Y., He, L. Q., Jiang, Y. J., Sun, M. M., Chen, E. J. and Lee, F. H. (2019). Effect of in situ water content variation on the spatial variation of strength of deep cement-mixed clay. *Géotechnique*, 69(5): 391–405.

Liu, Y., Jiang, Y. J., Xiao, H. and Lee, F. H. (2017). Determination of representative strength of deep cement-mixed clay from core strength data. *Géotechnique*, 67(4): 350–364.

Liu, Y., Lee, F. H., Quek, S. T., Chen, E. J. and Yi, J.-T. (2015). Effect of spatial variation of strength and modulus on the lateral compression response of cement-admixed clay slab. *Géotechnique*, 65(10): 851–865.

Liu, Y., Quek, S. T. and Lee, F. H. (2016). Translational random field with marginal beta distribution in modelling material properties. *Structural Safety*, 61: 57–66.

LTA (2020). *Engineering group materials & workmanship specification for civil & structural works, E/GD/09/104/A2, chapter 7—soil improvement works*, Land Transport Authority, Singapore.

Mackens, S., Fechner, T., Rios, N. I. C., Tweeton, D., Guerreros, J. C. G. and Niederleithinger, E. (2015). A new approach to determine the diameter of a jet grouted column using seismic methods. *Proc. Symp. Appl. Geophys. Eng. Env. Problems*: 142–146. https://doi.org/10.4133/SAGEEP.28-021.

Madhyannapu, R. S., Puppala, A. J., Nazarian, S. and Yuan, D. (2010). Quality assessment and quality control of deep soil mixing construction for stabilizing expansive subsoils. *Journal of Geotechnical and Geoenvironmental Engineering*, 136(1): 119–128. https://doi.org/10.1061/(ASCE)GT.1943-5606.0000188.

Melancon, J. L. and Shah, S. C. (1973). Soil-cement study—final report. *Research Report 72*, Research Project 68–9S, Louisiana Department of Highways.

Mongiovì, L., Croce, P. and Zaninetti, A. (1991). Analisi sperimentale di un intervento di consolidamento mediante gettiniezione. *Proc. II Convegno Nazionale dei Ricercatori del Gruppo di Coordinamento degli Studi di Ingegneria Geotecnica del C.N.R.*, Ravello, Italy: 101–118 (in Italian).

Navin, M. P. (2005). *Stability of embankments founded on soft soil improved with deep-mixing-method columns*. PhD thesis, VPI & State University, Blacksburg, VA, USA.

Ng, Y. C. H., Kek, H. Y., Hu, X., Wong, L. N., Teo, S., Ku, T. and Lee, F. H. (2022). An approach for characterising electrical conductivity of cement-admixed clays. *Soils and Foundations*, 62: 101127.

Ng, Y. C. H., Xiao, H., Armediaz, Y., Pan, Y. and Lee, F. H. (2020). Effect of short fibre reinforcement on the yielding behaviour of cement-admixed clay. *Soils and Foundations*. https://doi.org/10.1016/j.sandf.2020.03.009.

Nikbakhtan, B. and Osanloo, M. (2009). Effect of grout pressure and grout flow on soil physical and mechanical properties in jet grouting operations. *International Journal of Rock Mechanics and Mining Sciences*, 46(3): 498–505.

Osborne, N. H. and Ng, C. C. (2008). Strut omission by observational approach for deep excavation in Singapore using hybrid ground treatment. *Proc. Int. Conf. on Deep Excavation*, 10–12 November Singapore. Research Publishing Services (RPS), Singapore: 1–8

Phoon, K. K. and Kulhawy, F. H. (1999). Characterization of geotechnical variability. *Canadian Geotechnical Journal*, 36(4): 612–624.

Pieczyńska-Kozłowska, J. M. (2015). Comparison between two methods for estimating the vertical scale of fluctuation for modeling random geotechnical problems. *Studia Geotechnica et Mechanica*, 37(4): 95–103.

Puppala, A. J., Madhyannapu, R. S. and Nazarian, S. (2008). *Special specification for deep soil mixing. Product 0-5179-P1*, The University of Texas at Arlington.

Raito (2007). *RAS-JET system—mechanical soil mixing combined with jet grouting*. https://www.raito.co.jp/english/construction/pdf/ras-jet.pdf.

Shibazaki, M., Yoshida, H. and Matsumoto, Y. (1996). Development of a soil improvement method utilizing cross jet. In *Grouting and deep mixing* (Eds. R. Yonekura and M. Shibazaki). Balkema, Rotterdam, the Netherlands: 707–710.

Tatsuoka, F. (2010). Cement-mixed soils in Trans-Tokyo Bay Highway Project. *Soils and Foundations*, 50(6): 785–804.

Valdez-Llamas, Y. P., Auvinet, G. and Nunez, J. (2003). Spatial variability of the marine soil in the Gulf of Mexico. *Proc. OTC, Houston*, TX, OTC15266.

Vanmarcke, E. H. (1983). *Random fields: Analysis and synthesis*. MIT Press, Cambridge.

Western, A. W. and Grayson, R. B. (1998). The Tarrawarra data set: Soil moisture patterns, soil characteristics, and hydrological flux measurements. *Water Resources Research*, 34(10): 2765–2768.

Wong, D. Y. C., Ng, Y. N. C. H., Hong, Y., Ku, T., Goh, S. H. and Lee, F. H. (2025). Detection and stiffness measurement of weak zones in cement-treated ground using travel-time tomography. *Canadian Geotechnical Journal*, 62(2): 1–17.

Xiao, H., Lee, F. H. and Chin, K. G. (2014). Yielding of cement-treated marine clay. *Soils and Foundations*, 54(3): 488–501.

Yao, K., Chen, Q., Ho, J., Xiao, H. and Lee, F. H. (2018). Strain-dependent shear stiffness of cement-treated marine clay. *Journal of Materials in Civil Engineering*, 30(10): 04018255.

6 Designing with Chemically Improved Soils

6.1 DESIGN ISSUES

A primary result of deep soil mixing and jet-grouting processes is to create hardened columnar inclusions in the ground. These columnar inclusions posed challenges to conventional geotechnical analysis and design using manual calculations. In order to address these challenges, analytical and design methodologies were developed. These include the use of stress concentration ratios as well as guidelines for analysing the effects of columnar inclusions in soft soil. Many of these analytical methodologies were developed in the 1980s and 1990s and they are largely applicable to foundation and embankment stability design (e.g. Kitazume and Terashi 2013).

Several strands of development in the past two decades have significantly altered the scenario. Firstly, chemical stabilisation is increasingly used in other types of geotechnical construction, such as subgrade improvement in roadwork (e.g. Little and Nair 2009), temporary support in deep excavation and tunnelling (e.g. Shirlaw et al. 2000b) as well as mass stabilisation of marginal or contaminated soils (e.g. Thyagaraj et al. 2012). Subgrade improvement and mass stabilisation often involve mass, rather than, deep mixing. Mass mixing usually produces a contiguous mass of improved soil rather a columnar structure. In deep mixing for deep excavation and tunnelling, the deep-mixed columns are usually contiguous since transmission of lateral loads is a common objective. All these improved soil configurations differ significantly from the scenario of columnar inclusions separated by soft soil, which is typically encountered in foundation and embankment works.

Another strand of development is the advent of computational technologies such as numerical analysis software applications, which enable stress analysis to be conducted for much more complex configurations than was previously possible. In many instances, numerical codes have obviated the necessity to conduct hand calculations. Instead, new issues have arisen which relate to the selection of appropriate improved soil strength parameters for design that consider the spatial variability of the treated soil. In the following discussion, the analytical approaches for evaluating stress concentration ratio, settlement under one-dimensional loading and stability of embankment overlying ground reinforced by soil-cement columns will be summarised. Subsequent sections will focus on determination of representative improved soil properties considering spatial variability of the treated ground.

DOI: 10.1201/9781003285885-6

6.2 CHEMICALLY ADMIXED SOIL COLUMNS AS FOUNDATIONS

One of the early applications of cement-treated soil columns is as foundation elements underlying embankments, highways and earth-retaining structures. In such applications, the soil columns are usually installed in discrete, that is non-contiguous arrangements. The objective is similar to embankment piles, granular columns and sand compaction piles, that is to create stiff columnar inclusions in the otherwise soft soil. The analytical approach is also similar to that of granular columns and embankment piles and consists mainly of three aspects, namely overburden stress distribution, settlement and stability.

6.2.1 Stress Distribution Between Soft Soil and Columnar Inclusions

The stress distribution between soft soil and columnar inclusions is typically calculated using BS-8006's (2010) approach with Hewlett and Randolph's (1988) method or EBGEO's (2011) approach. The BS-8006 (2010) approach ignores the lateral support of the soft soil on the sides of the columnar inclusion; this is not expected to affect the results significantly as the treated soil columns are typically much stiffer and stronger than the soft soil. Hewlett and Randolph's (1988) method assumes that the soil arches in the embankment soil that redistribute stresses to the treated soil columns are roughly hemispherical domes. For square-shaped treated soil columns with width a, the arching-induced stress σ'_{col} on the top of the column is given by

$$\sigma'_{col} = \frac{(\gamma H + \sigma_{\text{sur}})S^2}{a^2}\frac{\beta}{1+\beta} \tag{6.1}$$

where γ and H are the unit weight and height of the embankment fill, respectively, σ_{sur} is the surcharge pressure on the embankment surface, S is the centreline spacing between treated soil columns and β is given by

$$\beta = \frac{2K_p}{(K_p+1)\left(1+\frac{a}{S}\right)}\left[\left(1-\frac{a}{S}\right)^{-K_p} - \left(1+\frac{a}{S}K_p\right)\right] \tag{6.2}$$

where $K_p = \dfrac{1+\sin\phi'}{1-\sin\phi'}$ is the Rankine passive earth pressure coefficient of the embankment soil.

The average vertical stress on the soil σ'_{soil} can be found by vertical equilibrium of the square unit cell comprising the treated soil column and surrounding soft soil, that is

$$\sigma'_{\text{col}}a^2 + \sigma'_{\text{soil}}\left(S^2 - a^2\right) = (\gamma H + \sigma_{\text{sur}})S^2 \tag{6.3}$$

Together with Equation (6.1), this gives

$$\sigma'_{\text{soil}} = \frac{(\gamma H + \sigma_{\text{sur}})S^2}{(S^2 - a^2)} \frac{1}{1+\beta} \tag{6.4}$$

EBGEO (2011) considers the lateral support from the soft soil on the treated soil column. The stress acting on the soft soil is given by

$$\sigma'_{\text{soil}} = \lambda_1^{\chi}\left(\gamma + \frac{\sigma_{\text{sur}}}{H}\right)\left\{h_g\left(\lambda_1 + h_g^2\lambda_2\right)^{-\chi} + h_g\left[\left(\lambda_1 + \frac{h_g^2\lambda_2}{4}\right)^{-\chi} - \left(\lambda_1 + h_g^2\lambda_2\right)^{-\chi}\right]\right\} \tag{6.5a}$$

$$\text{where } \lambda_1 = \frac{(S-a)^2}{8} \tag{6.5b}$$

$$\lambda_2 = \frac{S^2 + 2aS - a^2}{2S^2} \tag{6.5c}$$

$$\chi = \frac{a\left[\tan^2\left(45° + \frac{\phi'}{2}\right) - 1\right]}{\lambda_2 S} \tag{6.5d}$$

$$\text{and } h_g = \begin{cases} \text{S/2 for } H \geq \text{S/2} \\ H \text{ for } H < \text{S/2} \end{cases} \tag{6.5e}$$

The stress acting on the treated soil column can then be determined using Equation (6.3).

For a circular treated soil column with diameter d, an equivalent column width a_{eq} can be estimated by equating the area of an equivalent square section to that of the circular section, giving

$$a_{\text{eq}} = \frac{d}{2}\sqrt{\pi} \tag{6.6}$$

Broms (2004) also proposed a simple relationship for estimating the "stress factor," which is the ratio of the stress increment in the treated soil column to the overburden stress increment, based on an area-weighted equivalent constrained modulus M_{eq} of the treated ground, given by

$$M_{\text{eq}} = AE_{\text{col}} + (1 - A)M_{\text{soil}} \tag{6.7}$$

where A is the proportion of overall area that is occupied by treated soil columns, also known as area replacement ratio, E_{col} is the Young's modulus of the treated soil column and M_{soil} is the constrained modulus of the soft soil. The use of M_{soil} assumes that the soft soil will deform sufficiently in the lateral direction to mobilise the lateral constraint from the treated soil columns. On the other hand, the use of E_{col}

assumes that the treated soil column will not deform sufficiently in the lateral direction to mobilise any lateral constraint from the soft soil. Assuming that the vertical strain of the treated soil column is equal to that of the soft soil leads to

$$\sigma'_{col} = \frac{(\gamma H + \sigma_{sur})E_{col}}{\left[AE_{col} + (1-A)M_{soil}\right]} \tag{6.8}$$

It should be noted that Hewlett and Randolph's (1988) and EBGEO's (2011) relationships are based on limit equilibrium conditions, wherein the arching action in the embankment is fully mobilised. On the other hand, Broms' (2004) relationship assumes that vertical strain differential between treated columns and soft soil is negligible and arch action is not fully mobilised. The previously discussed relationships apply to scenarios wherein the treated soil columns are terminated into stiff soil layers. For "floating" treated soil columns, which are terminated within soft soil layers, the stress difference between treated soil columns and soft soil is likely to be much lower.

6.2.2 Bearing Capacity and Embankment Stability

The bearing capacity of a single treated soil column is governed by either failure of the soft soil surrounding the column or failure of the column itself. The bearing capacity of the soft soil surrounding the column is often governed by short-term (undrained) failure since the strength of the soft soil typically increases as it consolidated with time. The ultimate bearing load $Q_{ult,soil}$ of a single pile floating in soft soil may be estimated using Broms' (1991) relationship

$$Q_{ult,soil} = \left(\pi d H_{col} + 2.25\pi d^2\right)c_u \tag{6.9}$$

where H_{col} is the length of the treated soil column and c_u is the average undrained shear strength of the surrounding soft clay. This relationship assumes that the skin friction resistance at the interface between treated soil column and soft soil is equal to the undrained shear strength and that the point resistance of the soft soil is $9c_u$. For lime columns, when the undrained shear strength exceeds 30kPa, Broms (1991) recommended using a reduced shear strength to calculate the skin resistance.

The short-term ultimate bearing capacity, at depth z, of the treated soil column $Q_{ult,col}$ can be estimated using Broms (1991) relationship:

$$Q_{ult,col} = A_{col}\left(3.5c'_{col} + K_{p,col}\sigma_h\right) \tag{6.10}$$

where A_{col} is the cross-sectional area of the column, c'_{col} is the cohesion of the cement-treated soil, $K_{p,col}$ is the Rankine passive earth pressure that can be mobilised in the treated soil column and σ_h is the lateral earth pressure acting on the soil column at depth z.

Tyagi et al. (2017) cited observations from Chin (2006) and Xiao (2009) that pore water was expelled from the sides of specimens of cement-treated clay during

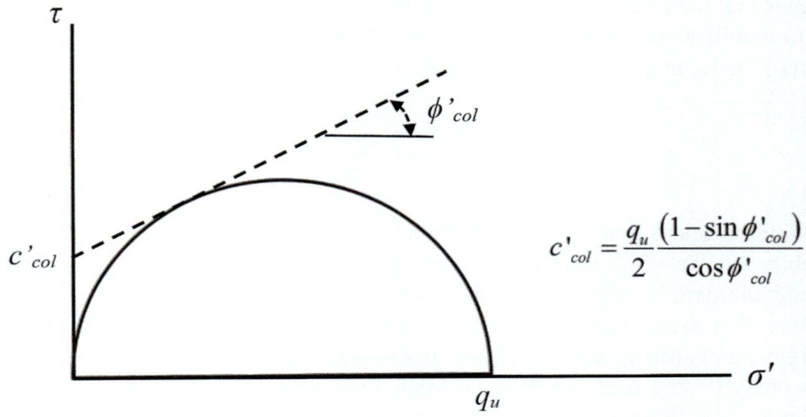

FIGURE 6.1 Relationship between cohesion and angle of friction of the treated soil and the unconfined compressive strength.

unconfined compression test. They postulated that, for cement-admixed soil with high cement content, an unconfined compression test is likely to take place under drained conditions, owing to the high stiffness of the improved soil. As Figure 6.1 shows, under drained conditions, the cohesion c'_{col} is related to the unconfined compressive strength q_u and the angle of friction of the treated soil ϕ'_{col} by

$$c'_{col} = \frac{q_u}{2}\frac{\left(1-\sin\phi'_{col}\right)}{\cos\phi'_{col}} \tag{6.11}$$

Broms (1991) assumed that $\phi'_{col} = 30°$, which leads to $K_{p,col} = 3$ and $c'_{col} \sim 0.29q_u$. In reality, ϕ'_{col} may differ significantly from 30°. Broms (2004) proposed an angle of friction of 40° for cement-treated columns. Suzuki's (1982) data for cement-treated soil from Japan with 5% to 15% cement content indicate that the angle of friction ranges from 25° to 35°, with c'_{col} ranging from $0.27q_u$ to $0.37q_u$. Xiao's (2009) data from consolidated undrained triaxial tests using cement-treated marine clay specimens with 10% to 50% cement content give critical state angles of friction of between 36° to 42°. Tyagi et al. (2017) attributed the higher angle of friction obtained by Xiao (2009), compared to that of Suzuki (1982), to the higher cement content. If this is so, then the angle of friction is likely to vary with binder content.

For the stability of column groups supporting superstructures, Broms (1991) proposed that the column group and the intervening soil can be analysed as a foundation block. This leads to the undrained bearing capacity $Q_{ult,grp}$ of the column group being the sum of the skin friction and end bearing of the group, that is

$$Q_{ult,grp} = 2c_{u,side}H_{col}\left(B+L\right)+N_c c_{u,base}BL \tag{6.12}$$

where B and L are the length and width of the column group footprint and $c_{u,side}$ and $c_{u,base}$ are the undrained shear strengths of the soil on the side and base of the group.

Equation (6.12) only applies to external instability of the column group. For local failure, Broms (1991) proposed using a relationship based on a circular slip surface passing through part of the column group.

However, Kitazume et al.'s (1996) and Karastanev et al.'s (1997) centrifuge test results indicate that internal failure of the column group can also be caused by columns near the centreline of group failing by compressive shear and those nearer the periphery failing by bending-induced brittle tensile rupture. They also noted that the columns fail one by one in a progressive failure manner, rather than simultaneously. This indicates that internal failure of column groups is much more complex than circular slip failure.

Broms (2004) proposed relationships for assessing the capacity of treated soil columns under combined lateral and vertical loading, assuming that the undrained shear strength of the surrounding soil is fully mobilised. However, this may not always be true as the soft soil will typically require a much large shear strain to mobilise its full shear strength than that required for the column to mobilise its bending capacity. Furthermore, since the treated soil columns fail progressively and in a brittle manner, the full bending capacity of all the columns will not be mobilised simultaneously. Instead, individual columns will rupture progressively and those which have ruptured will likely lose most, if not all, bending capacity. Hence, a limit equilibrium approach based on simultaneous full mobilisation of column and soil capacities is unlikely to be able to reflect the failure behaviour. Slope stability analysis may also not be applicable as it also assumes that column and soil capacities are fully mobilised simultaneously. In addition, slope stability analysis considers only shear failure, not bending failure.

Progressive failure may be better analysed using numerical algorithms such finite element or finite difference provided an appropriate material model can be prescribed. Such a material model will need to have features which account for tensile rupture, such as tensile strength and tensile softening (e.g. Xiao et al. 2017).

6.2.3 Settlement of Admixed Column Groups

The settlement of an admixed column group is usually calculated by assuming that the axial compression of the columns is the same as the vertical compression of the intervening soft soil. This assumption is adequate if the spacing of the column is less than 1m to 1.5m (e.g. Liedberg et al. 1996a, 1966b). In such cases, Broms (2004) proposed that the settlement of the column group and intervening soft soil can be estimated based on the area-weighted modulus of the column group and soft soil, that is Equation (6.8). With this, the vertical compression of the column group s_{col} is given by

$$s_{col} = \frac{\left(\gamma H + \sigma_{sur}\right) H_{col}}{\left[A E_{col} + \left(1 - A\right) M_{soil}\right]} \tag{6.13}$$

Settlement of the improved column group which is terminated into stiff bearing layer is usually smaller as long as stress redistribution action of the overlying embankment or soil is adequate. In some cases, admixed soil columns are terminated within compressible soil layer. Settlement of such floating column groups must then include the compression of the compressible soil beneath the column group.

6.3 CONTIGUOUS COLUMNAR STRUCTURES

Contiguous columnar structures used in retaining or stabilising structures in stability of slopes and excavations are typically analysed using numerical analyses such as finite element and finite differences. In cases where there is unstabilised soil in between columns, equivalent properties may be specified to avoid modelling the individual columns. Broms (2004) proposed a relationship for the equivalent shear strength f_{eq} along the slip surface, based on weighting the strength parameters by the area replacement ratio A, which is given by

$$f_{eq} = A\left(\sigma'_{f,col}\tan\phi'_{col} + c'_{col}\right) + \left(1 - A\right)\left(\sigma'_{f,soil}\tan\phi'_{soil} + c'_{soil}\right) \tag{6.14}$$

where $\sigma'_{f,col}$ and $\sigma'_{f,soil}$ are the effective normal stresses on the columns and soil, respectively, along the rupture surface, ϕ'_{col} and ϕ'_{soil} are the angles of friction of the column and soil, respectively, and c'_{col} and c'_{soil} are the effective cohesion of the treated columns and soil, respectively. If the treated soil columns and unstabilised soil are both in an undrained condition, then the angles of friction of the soil and column vanish and cohesions of the soil and column become the respective undrained shear strength, that is $\tan\phi'_{col} = \tan\phi'_{soil} = 0°$, $c'_{col} = c_{u,col}$ and $c'_{soil} = c_{u,soil}$, giving

$$f_{eq} = Ac_{u,col} + \left(1 - A\right)c_{u,soil} \tag{6.15}$$

In the design of earth-retaining and stabilising structures such as cement-admixed soil grid and block retaining structures (e.g. Yee and Tan 2015) and cement-admixed soil slabs (e.g. Liu et al. 2015) in Singapore and Malaysia, Equation (6.15) is typically used since the treated and untreated soils are both taken to be in an undrained state, but with $A = 1$. This is because the cement-admixed columns are required to transmit lateral compressive stresses. Hence, the columns are often constructed so that they overlap one another. If there is no error in column positioning, then the amount of untreated soil in between columns is minimal. The undrained shear strength of the treated soil $c_{u,col}$ is typically taken to be half of the unconfined compressive strength. This implicitly assumes that the unconfined compression test is in an undrained condition. It should be noted that Equations (6.14) and (6.15) do not consider spatial variation in the properties of the treated columns. Methods of considering spatial variability will be discussed later. In addition, Equation (6.14) only applies to the conditions on a potential slip surface.

6.4 ACCOUNTING FOR SPATIAL VARIATION IN SOIL PROPERTIES

6.4.1 Current Practice: Minimum Strength with Respect to Design Strength

The primary effect of spatial variation in soil properties, such as its strength, is to alter the overall performance of the treated soil domain. Conventional geotechnical analyses such as limit equilibrium and continuum analyses such as finite element and finite difference do not address spatial variation in strength. This is because the respective material zones, such as soft clay or treated soil, are each typically considered as a uniform material. In recent years, a significant amount of work on random finite element analyses has been reported, some of which will be discussed shortly. However, random finite element analysis remains relatively uncommon in geotechnical design, owing to the significantly higher demand on computational resources and time, compared to deterministic analysis.

Instead, the effects of spatial variation are dealt with by adopting a conservative value for strength (and modulus) in the design process and ensuring that this value is exceeded or equalled by the strength (and modulus) of all or a substantial proportion of cored samples. For instance, Puppala et al. (2008) stipulated a 14- and 28-day minimum unconfined compressive strength of 100psi and 150psi, respectively. Similarly, for cement–soil treatment related to underground construction projects in Singapore, a minimum undrained shear strength, taken to be half of the unconfined compressive strength, of about 400kPa is often stipulated. For instance, in the construction of the Marina South Station and tunnels of the Thomson-East Cost Line in Singapore, specifications on chemical improvement works to mitigate ongoing consolidation of the soft marine clay include requirements that the undrained shear strength and Young's modulus of the treated soil shall not be less than 450kPa and 190MPa, respectively. For other regions around the tunnels to be dug, higher strength and modulus were specified.

A variation of this approach permits a certain percentage of the core samples to have strength lower than the design strength. This is often subjected to the strength of all core samples exceeding a minimum value that is lower than the design strength. For example, EPA (2012) stipulated the following:

> The soft deposits shall be mixed and treated as one unit which shall achieve the following specifications after 28-day cure:
>
> - Unconfined compressive strength, q_u, of 50 psi or greater for all treated cells or columns;
> - Permeability of less than 5 x 10^{-6} cm/sec for all treated cells or columns.
> - No more than 10% of the tests may be less than the design strength value.

Similarly, in the ground improvement work for the construction of the Tanjong Rhu Station of the Thomson-East-Coast Line in Singapore, the design undrained shear strength is 800kPa. As Table 6.1 shows, the required minimum undrained shear strength and modulus is 50% of the design strength and modulus, respectively. The required mean strength is about 29% higher than the design strength for sample sizes not exceeding 50. For sample sizes exceeding 50, the required mean strength is 12.5% higher than the design strength. The smaller difference between the mean and the design strengths for sample sizes exceeding 50 underscores the fact that larger samples are intrinsically more representative of the population than smaller samples. A similar approach is used for ground improvement work for the Changi Airport Terminal 5 substructure, where the minimum required strength is 50% of the design strength and the mean strength is 1.25 times the design strength.

TABLE 6.1
Requirements for ground improvement for the Tanjong Rhu Station, Thomson-East Coast Line, Singapore.

	Sample size: 10[1] to 50		Sample size: >50	
	Undrained shear strength (kPa)	Undrained modulus (MPa)	Undrained shear strength (kPa)	Undrained modulus (MPa)
Design	800	280	800	280
Minimum	400	140	400	140
Mean[2]	1030	350	900	310

1 Minimum sample size is 10.

2 For calculation of the mean values, undrained shear strength and modulus higher than 4000kPa and 1500MPa, respectively, are to be capped at these values.

6.4.2 Statistical Approach and Characteristic Strength

The previous sub-section underscores the progressive migration from a minimum specified strength, which must be equalled or exceeded by all samples, hereafter termed "sample minima" approach, to a more "statistical" approach based on a minimum percentage of samples satisfying a specified or design strength. Bruce et al. (2013, p. 187) recommended that "90% of all samples across the site should equal or exceed the specified strength"; this corresponding to a 90% confidence level. This confidence level is slightly lower that the 95% confidence level recommended by Eurocode EC7 (BS EN 1997–1 2004).

Topolnicki (2016) proposed the use of a characteristic strength, similar to that used by BS EN 1997–1 (2004), based on the assumption that the compressive strength of the core samples following a log-normal distribution. The characteristic strength, as specified by BS EN 1997–1 (2004), is a cautious estimate of the strength of the soil that addresses strength variability and thus uncertainty in the strength to be used in design. BS EN 1997–1 (2004) recommends the use of 95% confidence level, which is a value that has a 95% chance of being exceeded. For scenarios where there is a strong averaging effect in the geotechnical system, BS EN 1997–1 (2004) permits the use of the 95% confidence level mean estimate as the characteristic core strength q_{ck}. This is the mean estimate which has a 95% chance of exceedance. If the strength follows a normal distribution, the characteristic strength is then given by

$$q_{ck} = \mu(q_c) - m\frac{S_d(q_c)}{\sqrt{n}} \tag{6.15}$$

where n is the sample size and m is an "acceptance factor" (BS EN 12716 2018) which is used to adjust the confidence level and thus the degree of conservatism. For 95% confidence level, $m = 1.645$. If there is no averaging effect in the geotechnical system, then $n = 1$ and q_{ck} reduces to the 5% fractile of the sample data, that is the value which is exceeded by 95% of the sample.

The strength of deep-mixed or jet-grouted soil often follows a log-normal distribution. This is a probability distribution of a random variable, that is the strength of the core samples q_c, such that $\ln(q_c)$ follows a normal distribution. The probability distribution $p(q_c)$ of the log-normal distribution is given by

$$p(q_c) = \frac{1}{q_c S_d(\ln q_c)\sqrt{2\pi}} \exp\left[-\frac{\left(\ln q_c - \mu(\ln q_c)\right)}{2\left(S_d(\ln q_c)\right)^2}\right] \tag{6.16}$$

where $\mu(\ln q_c) = \frac{1}{n}\sum_{i=1}^{n} \ln(q_{ci})$ and $S_d(\ln q_c) = \sqrt{\frac{1}{n-1}\sum_{i=1}^{n}\left[\ln(q_{ci}) - \mu(\ln q_{ci})\right]^2}$ are the mean and standard deviation of the natural logarithm of random variable, and q_{ci} denotes the ith core strength value. The characteristic compressive strength q_{ck} is then given by

$$q_{ck} = \exp\left[\mu(\ln q_c) - mS_d(\ln q_c)\right] \tag{6.17}$$

Equation (6.17) assumes that $n = 1$ [see Equation (6.15)]; that is, there is no averaging effect. For confidence levels of 80%, 85%, 90% and 95%, m = 0.842, 1.037, 1.282 and 1.645, respectively. Topolnicki (2016) also recommended an upper limit of 12MPa for the characteristic core strength q_{ck}, which is unlikely to be exceeded in most realistic scenarios.

Topolnicki (2016) proposed that, where deep mixing is employed to stabilise a geotechnical system which would otherwise be unstable, the confidence level should be 90% (m = 1.282). For systems which would be stable even without ground treatment and the latter is used to satisfy the serviceability limit state (SLS) requirement, such as limiting ground movement, he proposed that the confidence level should be 80% to 85%, depending upon the criticality of the treatment work to the SLS requirement. It should be noted, however, that these confidence levels are all lower than the recommended value of 95% in BS EN 1997–1 (2004).

As discussed in the previous sub-section, statistical approaches have been used in practice. The characteristic strength has also been used in some projects in Singapore. One example is the construction of the Changi Airport Terminal 5 substructure, where an alternative ground improvement criterion is for the 90%-confident characteristic mean value determined from a minimum of ten samples to be greater than the design value. Nonetheless, in many other instances, the specified strength which is required to be achieved may still be quite conservative. For instance, Yohannes and Daramalingam (2019) used this framework to assess the level of conservatism in seven project sites in Malaysia using sample minima as a criterion. They noted that, in all cases, the specified strength which is required to be equalled or exceeded by the core strength corresponds to a confidence level exceeding 90%. In several instances, the confidence level exceeded 99%.

6.4.3 Reliability Analysis

6.4.3.1 Brief Overview

While the statistical approach discussed in the last sub-section is an improvement on the sample minima approach, it is still incomplete as a design procedure, since the characteristic strength is used therein as a quality control measure, rather than as a design parameter. There are also differences when compared with the "material mobilisation factor" approach of BS EN 1997–1 (2004), which factors the characteristic strength further downwards by mobilisation factors, typically 1.25 for c' and ϕ' and 1.4 for c_u to obtain the design strength. Hence, in the BS EN 1997–1 (2004) "material mobilisation factor" approach, the design strength is lower than the characteristic strength. One can think of the characteristic strength as a way of accounting for variability, while the design strength, which incorporates mobilisation and other partial factors, is used to ensure reserve capacity and adequate margin (or factor) of safety against the limit state. This issue does not arise in the other BS EN 1997–1 (2004) approaches, wherein the partial factors are applied to the loading rather than the material strength. The acceptable probability of failure in permanent geotechnical structures is typically less than 0.01. For instance, in ANCOLD (1998), the highest acceptable annual probability of failure is 0.001. This is much lower than the 5% fractile level typically used for characteristic strength. Hence, on its own, the characteristic strength is deemed to have insufficient margin or factor of safety against the limit state, unless it is used together with load factoring. This leads to an over-arching question: How does the specified strength relate to overall geotechnical system performance and margin of safety?

Reliability analysis has been widely used to analyse systems with inherent uncertainties in the input parameters. The following discussion provides only a brief overview of reliability analysis and is not intended to supplant the many good textbooks on reliability analysis (e.g. Baecher and Christian 2003). The underlying premise in reliability analysis is that the uncertainties in the input parameters can potentially affect the computed factor of safety. Hence the factor of safety is a statistical, rather than a deterministic, parameter. One can think of the parameters as consisting of two groups, loading Q_L and resistance R_S, both of which are statistical, rather than deterministic parameters. The action of Q_L is to bring the system closer to the limit state whereas the action of R_S is to move the system away from the limit state. The loading Q_L includes, but is not limited to, the self-weight, overturning earth pressure and surcharge loading. The resistance R_S may include material resistance or strength and restoring forces. Although not always corrected, it is common to assume that R_S and Q_L are normally distributed with positive means μ_R and μ_Q, and variances σ^2_R and σ^2_Q. Then the margin of safety is defined as

$$M = R_S - Q_L \tag{6.22}$$

If R_S and Q_L are both normally distributed, as illustrated in Figure 6.2, then M will also be normally distributed, with the mean μ_M and variance σ^2_M given by

$$\mu_M = \mu_R - \mu_Q \tag{6.23}$$

and

$$\sigma_M^2 = \sigma_R^2 + \sigma_Q^2 - 2\rho_{RQ}\sigma_R\sigma_Q \tag{6.24}$$

The resistance R_S and loading Q_L are usually taken to be uncorrelated. In such situations, their correlation coefficient $\rho_{RQ} = 0$, and Equation (6.24) reduces to

$$\sigma_M^2 = \sigma_R^2 + \sigma_Q^2 \tag{6.25}$$

We can then define a reliability index β as

$$\beta = \frac{\mu_M}{\sigma_M} = \frac{\mu_R - \mu_Q}{\sqrt{\sigma_R^2 + \sigma_Q^2}} \tag{6.26}$$

The reliability index β quantifies the distance of the mean margin of safety from the limit state, in terms of the number of standard deviations. It is also the reciprocal of the coefficient of variation of M. As Figure 6.2 illustrates, the probability of failure is the probability that $M < 0.0$.

In geotechnical engineering, the factor of safety is more widely used than the margin of safety. The factor of safety F is defined by

$$F = \frac{R_S}{Q_L} \tag{6.27}$$

The limit state is reached when $F = 1$. Hence, a reliability index β that expresses the distance of the factor of safety from the limit state can be defined as

$$\beta = \frac{\mu_F - 1}{\sigma_F} \tag{6.28}$$

where μ_F and σ_F are the mean and standard deviation of F (Figure 6.3).

However, calculation of μ_F and σ_F is generally more difficult than that of μ_M and σ_M. This is because F is a ratio of two statistical variables and the probability distribution of F is a ratio distribution (e.g. Díaz-Francés and Rubio 2013). Even if R_S and Q_L are normally distributed, the distribution of F, in general, is non-normal, is

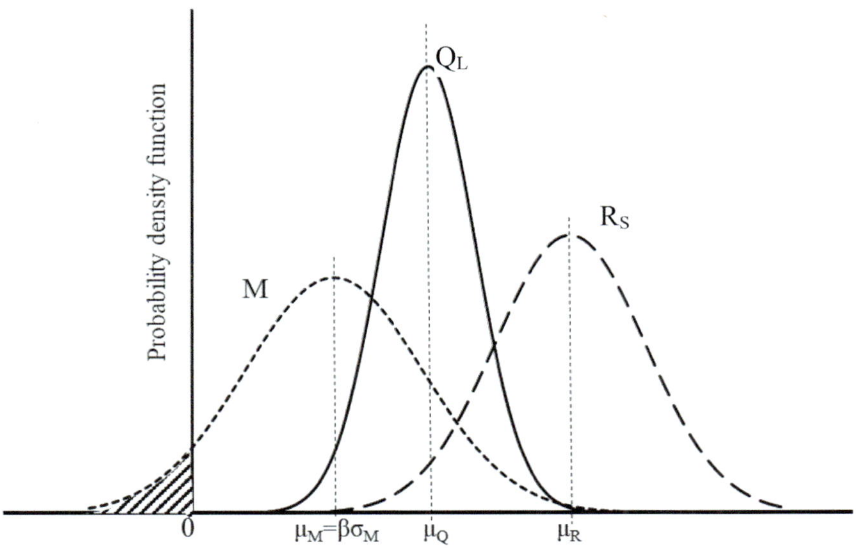

FIGURE 6.2 Probability density functions loading (Q_L), resistance (R_S) and margin of safety (M). Area of shaded region represents the probability of failure.

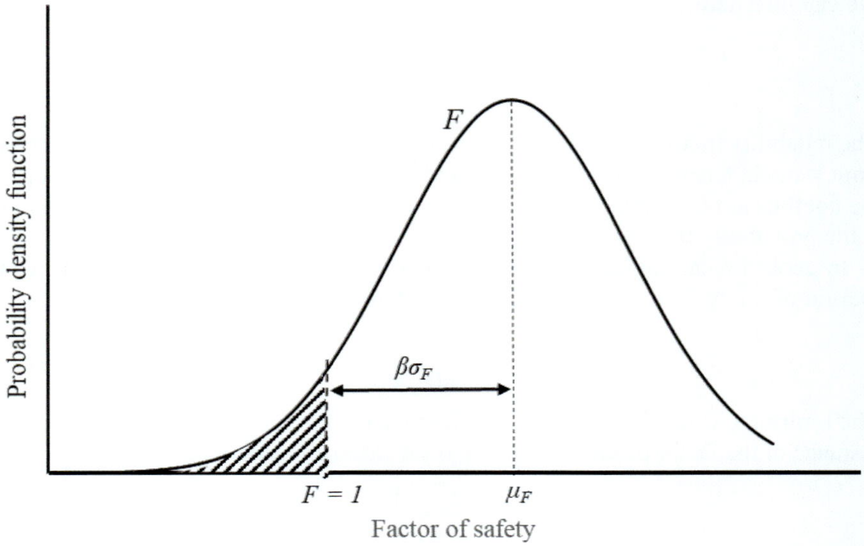

FIGURE 6.3 Relationship between the mean and standard deviation of the factor of safety and the reliability index.

positively skewed, is heavy-tailed and has no stable moment (such as the mean) as the number of samples tends to infinity. The "tailedness" of a distribution describes the relative proportion of outlier values, and a heavy-tailed distribution is one with relatively large numbers of outlier values. The absence of a stable mean implies that, as the number of samples tends to infinity, the sample mean does not converge to a stable population mean. Higher-order moments such as the variance tend to diverge even faster. However, using numerical examples, Díaz-Francés and Rubio (2013) demonstrated that the ratio distribution of F can still be reasonably approximated by the normal distribution under the following conditions:

1. The means of the loading Q_L and resistance R_S are both positive. This is usually the case in the majority of geotechnical scenarios.
2. The coefficient of variation of R_S, that is the numerator term, is less than 1.0. This is also usually met in most geotechnical scenarios.
3. The coefficient of variation of Q_L, that is the denominator term, is less than 0.25. For coefficient of variation of Q_L between 0.25 and 0.3, the distribution of F becomes increasingly left-skewed although, from an engineering viewpoint, the discrepancy may still be acceptable.

The absence of stable moments may also not be a critical issue if one is dealing with a finite number of samples rather than a population of infinite samples. Thus, despite the mathematical objections, a normal distribution is still assumed for F in many instances (e.g. Baecher and Christian 2003; Filz and Navin 2006). With this assumption, the probability of failure p_f is related to the reliability index β by

$$p_f = 1 - \frac{1}{\sqrt{2\pi}} e^{-\frac{\beta^2}{2}} \tag{6.29}$$

Hence, the probability of failure can be evaluated once the reliability index β is determined.

Many approaches have been proposed for evaluating the reliability index (e.g. Baecher and Christian 2003). The following discussion provides a brief overview of some commonly used methods.

6.4.3.2 Methods of Evaluation

The factor of safety F [Equation (6.27)] is typically represented as the quotient of the resistance R_S divided by the loading Q_L. In reality, both R_S and Q_L are multivariate sets; that is R_S consists of multiple resistance items [r_1, r_2, r_3,. . . . . . . r_m] while Q_L consists of multiple loading items [q_1, q_2, q_3,. . . . . . . q_n]. Each of the resistance items has a mean value which collectively can be represented by [μ_{R1}, μ_{R2}, μ_{R3},. . . . . . .μ_{Rm}]. Similarly, each of the loading items has a set of mean values [μ_{Q1}, μ_{Q2}, μ_{Q3},. . . . . . .μ_{Qn}]. In general, if the system lies within the safe zone, then F calculated using the mean values of the resistance and loading item should be greater than 1.0. Certain combinations of values of [r_1, r_2, r_3,. . . . . . . r_m] and [q_1, q_2, q_3,. . . . . . . q_n] will bring the system to a limit state, that is $F = 1.0$. These combinations together form the "performance surface" of R_S and Q_L. The scenario is illustrated using the "performance space" in Figure 6.4. For simplicity, only one item each from R_S and

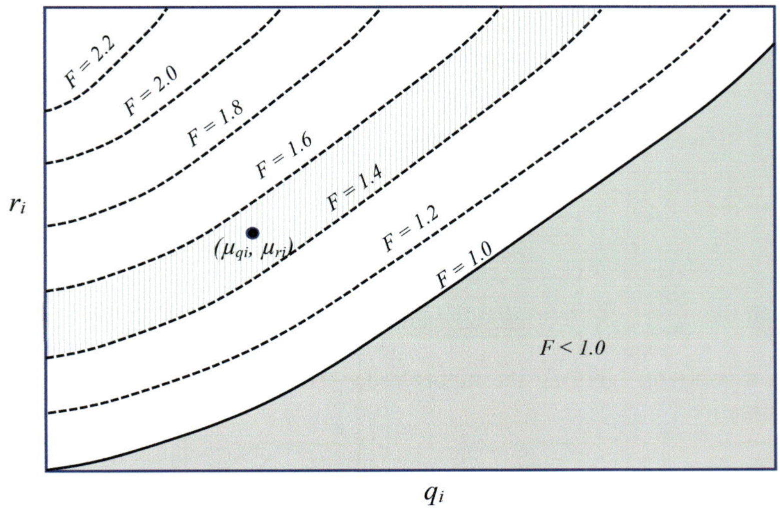

FIGURE 6.4 Performance space with contours of factor of safety (F) for different combinations of resistance (r_i) and loading (q_i) parameters.

Q_L are depicted in this two-dimensional plot. In reality, the performance surface is a hypersurface in an $(m+n)$-dimensional performance hyperspace.

Figure 6.4 also shows the contours of the factor of safety corresponding to the different combinations of $[r_1, r_2, r_3, \ldots\ldots r_m]$ and $[q_1, q_2, q_3, \ldots\ldots q_n]$. The black point represents the mean values of $[r_1, r_2, r_3, \ldots\ldots r_m]$ and $[q_1, q_2, q_3, \ldots\ldots q_n]$, that is $[\mu_{r1}, \mu_{r2}, \mu_{r3}, \ldots\ldots \mu_{rm}]$ and $[\mu_{q1}, \mu_{q2}, \mu_{q3}, \ldots\ldots \mu_{qn}]$, which, in this illustration, gives a factor of safety of ~1.5. Combinations of $[r_1, r_2, r_3, \ldots\ldots r_m]$ and $[q_1, q_2, q_3, \ldots\ldots q_n]$ that result in $F < 1.0$ represent the unsafe region (shaded zone). The probability of failure is the joint probability of all the combinations of $[r_1, r_2, r_3, \ldots\ldots r_m]$ and $[q_1, q_2, q_3, \ldots\ldots q_n]$ that result in $F < 1.0$, that is the combinations which fall within the grey zone. Hence, the probability of failure can be determined by integrating the probability density function values of all the combinations of $[r_1, r_2, r_3, \ldots\ldots r_m]$ and $[q_1, q_2, q_3, \ldots\ldots q_n]$ within the grey zone. Similarly the probability of the factor of safety F lying between, say, 1.4 and 1.6 (the vertically hatched region), can be determined by integrating the probability density function values of all the combinations of $[r_1, r_2, r_3, \ldots\ldots r_m]$ and $[q_1, q_2, q_3, \ldots\ldots q_n]$ within this region.

Very often, the function of performance surface cannot be explicitly evaluated, and a numerical approach has to be employed. One possible approach is to use Monte Carlo simulation (Baecher and Christian 2003). More often, simplified approximate methods are used in place of direct integration, such as First Order Second Moment method (FOSM) (also known as Taylor Series method) (Ang and Tang 1975, 1984), Point Estimate method (Rosenblueth 1975) and Hasofer-Lind method, also known as First-Order Reliability method (FORM) (Hasofer and Lind 1974).

The FOSM or Taylor Series method is based on the truncation of the Taylor Series of the factor of safety F. As previously mentioned, F is a function of $[r_1, r_2, r_3, \ldots\ldots r_m]$ and $[q_1, q_2, q_3, \ldots\ldots q_n]$. For convenience, we will now lump the r_i's and q_i's together as $[x_1, x_2, x_3, \ldots\ldots x_k]$, where $k = m + n$. Suppose that we know the value of F at the mean values of all the x_i's and we want to express the value of F for other values of x_i's, we can express this in the form of a Taylor Series expansion, so that

$$F(x_1, x_2, \ldots\ldots x_k) = F(\mu_{x1}, \mu_{x2}, \ldots\ldots \mu_{xk}) + \sum_{i=1}^{k}\left(x_i - \mu_{xi}\right)\left.\frac{\partial F}{\partial x_i}\right|_{\mu_{xi}} + \text{higher order terms} \quad (6.30)$$

in which the $\left.\frac{\partial F}{\partial x_i}\right|_{\mu_{xi}}$ denotes that the partial derivative is evaluated at $x_i = \mu_{xi}$ and is thus a constant. Ignoring the higher order terms leads to the following linear approximation:

$$F(x_1, x_2, \ldots\ldots x_k) \approx F(\mu_{x1}, \mu_{x2}, \ldots\ldots \mu_{xk}) + \sum_{i=1}^{k}\left(x_i - \mu_{xi}\right)\left.\frac{\partial F}{\partial x_i}\right|_{\mu_{xi}} \quad (6.31)$$

The mean factor of safety μ_F is defined by

$$\mu_F = \iiint \ldots \int F\left(x_1..x_i..x_k\right) f_x\left(x_1...x_k\right) dx_1...dx_k \quad (6.32)$$

where $f_x(x_1 \ldots x_k)$ is the joint probability density function of the factor of safety. However, as Equation (6.31) shows, F consists of a constant $F(\mu_{x1}, \mu_{x2}, \ldots\ldots \mu_{xk})$ and a linear combination of single-variate terms. Since the mean of a constant is the constant itself, Equation (6.32) can be simplified to

$$\mu_F = F\left(\mu_{x1} \ldots \mu_{xk}\right) + \sum_{i=1}^{k} \left\{ \left. \frac{\partial F}{\partial x_i} \right|_{\mu_{xi}} \int_{-\infty}^{\infty} \left[\left(x_i - \mu_{xi}\right) f_x\left(\mu_{x1}, \ldots x_i \ldots \mu_{xk}\right)\right] dx_i \right\} \quad (6.33)$$

We noted that the partial derivative can be taken out of the integration process as it is a constant. Furthermore, since each term within the summation operation is single-variate, the k-dimensional integration reduces to a series of 1-dimensional integration. Finally, each integration operation involves evaluation of the first-moment of the probability density function around the mean value, which must be zero. Hence,

$$\mu_F = F\left(\mu_{x1}, \mu_{x12} \ldots \mu_{xk}\right) \quad (6.34)$$

That is, the mean factor of safety is also approximately the factor of safety of the mean values of the resistance and loading parameters. This is only true of the approximate equation, that is Equation (6.31), which ignores the higher-order terms. Using a similar approach, the standard deviation of the F can be expressed as

$$S_{dF} \approx \sqrt{\left(S_{dx1} \frac{\partial F}{\partial x_1}\right)^2 + \left(S_{dx2} \frac{\partial F}{\partial x_2}\right)^2 \ldots \left(S_{dxk} \frac{\partial F}{\partial x_k}\right)^2} \quad (6.35)$$

The values of each of the term can be evaluated by calculating the factor of safety at two known values of x_i. For example, if we evaluate F at $x_1 = \mu_{x1} + S_{dx1}$ and $_{x1} = \mu_{x1} - S_{dx1}$, then the difference in the factor of safety ΔF_1 represents the change of F over $\Delta x_1 = 2Sdx_1$, and the term $\left(S_{dx1} \frac{\partial F}{\partial x_1}\right)$ is then $\frac{\Delta F_1}{2}$. By changing each parameter in a similar manner, S_{dF} can be evaluated and the reliability index β can be determined using Equation (6.28).

The Point Estimate method and Hasofer-Lind method are both variants of the Taylor Series method. The Point Estimate method also involves evaluating the factor of safety at discrete points within the performance space, typically at (mean + 1 standard deviation) and (mean − 1 standard deviation). The Hasofer-Lind method is a refinement of the Taylor Series method, which involves an additional step of transforming each parameter to standard normal variate so that each parameter has a mean of 0 and a standard deviation of 1. This leads to a more robust estimate of the reliability index β.

6.4.3.3 Application to Chemically Treated Soil

There have been many instances of reliability analyses being applied to chemically treated soil. Filz and Navin (2006) applied reliability analysis to assess the stability of embankment supported by cement-treated soil columns installed in soft clay. They noted that results from the Hasofer-Lind method is in better agreement with the direct integration method than the Taylor Series and Point Estimate methods. It should be noted that the main target of reliability analysis is parametric uncertainty, rather than spatial variability. In conventional reliability, the parameters of a material zone, such as cement-treated soil, is taken to be uncertain, but uniform. This corresponds to a scenario in which the SoF is infinity. In order to deal with spatial variation, Filz and Navin (2006) used El-Ramly et al.'s (2002) approach of sub-dividing the cement-treated soil zone into different sub-zones, with dimensions equal to the SoF. Within each sub-zone, properties are still uniform. This creates a "block-like" spatial variation, which is ad hoc rather than random.

Al-Naqshabandy and Larsson (2013) also used reliability-based slope stability analysis to analyse embankments supported by lime-cement columns installed in soft soil. The treated soil domain was modelled as a composite material rather than as individual columns. The potential slip surface was sub-divided into numerous segments, each of which corresponds to a slice in the sliding portion of the slope. Each segment of the slip surface also has a local average shear strength and strength variance. The latter was related to the SoFs in the three directions as well as length and inclination of each segment, using Knabe et al.'s (1998) formulation. By summing up the local average and variance of the shear resistance along all the segment, the average shear resistance and its variance along the entire slip surface can be estimated. This allows the slip surface to be treated as a probabilistic entity, the properties of which are related to the CV and SoF of the composite soil.

In general, spatial variation is difficult to account for within the framework of conventional reliability analysis. This is especially so in non-limit equilibrium, continuum analysis such as finite element analysis, which requires spatial variation in constitutive behaviour to be considered, not just the shear resistance along the slip surface. In such cases, random finite element analysis, which is capable of simulating spatial variation in properties realistically, is now becoming much more widely used.

6.5 RANDOM FINITE ELEMENT ANALYSIS

6.5.1 Overview

The random finite element (RFE) method has been gaining popularity as a computational approach for studying the effects of spatial variation in soil properties. Early use of RFE method revolved around slopes and foundations in natural soils as well as seepage and settlement analysis (Griffiths and Fenton 2007). Early application to cement-treated soil includes axial capacity of treated soil column (e.g. Namikawa and Koseki 2013; Pan et al. 2018a), underground treated slab for deep excavation (e.g. Liu et al. 2015; Pan et al. 2018b) and tunnelling (e.g. Tyagi et al. 2018, 2020; Pan and Lee 2019; Pan et al. 2019). The basis of random finite element analysis is to simulate the random variation in properties in each element or integration point within a finite element, based on the mean, CV and SoF of the material property in question. An RFE analysis typically consists of the following steps:

(a) Generate a random field of required material properties in the material zone of the finite element domain. Typical random properties include the unconfined compressive strength, binder content or some other parameters of the constitutive model used. In general, there is no limit on the number of parameters which can be randomised. However, the random field generation and computational workload increases rapidly with each additional random variate. Hence, univariate random fields are most commonly employed. Each generation gives a random realisation of the spatially variable quantity in the material zone.

(b) Conduct a finite element analysis and evaluate the quantity of interest for the generated realisation.

(c) Regenerate the realisations and finite element analyses until the statistical properties, such as the mean and coefficient of variation, of the quantity of interest reach reasonably stable values. For instance, if the quantity of interest is the bearing capacity of a footing, then the realisation-analysis cycles are repeated until the mean and coefficient of variation of the computed bearing capacity reach reasonably stable values. In general, it typically requires a minimum of about 100 to 200 realisations to achieve stability, but this is not a hard-and-fast rule.

(d) The probability distribution of the quantity of interest provides the necessary statistical information on the effect of spatial variation. For example, the lowest 5th percentile of the results gives the bearing capacity that is exceeded approximately 95% of the time.

(e) A common item of interest is the equivalent uniform field, which will give the required level of performance as the spatially variable field. For example, if the random variable is the unconfined compressive strength and the performance is measured by the bearing capacity of a group of treated soil columns, one may be interested to determine the equivalent unconfined compressive strength which will, in a uniform material zone, give the same bearing capacity as the lowest 5% (of some other prescribed level) of the bearing capacity distribution. This can be readily determined by deterministic finite element analyses using uniform field, either directly in simple cases or by trial-and-error in more complicated cases.

6.5.2 Random Field Generation

There are many well-established methods of generating random fields (e.g. Fenton and Griffiths 2007). Common methods include spectral representation method (SRM) method, turning bands method (TBM) and local average subdivision (LAS) method. Some of these are discussed later. For a more complete discussion on methods of generating random field, readers are referred to literature relevant to this area (e.g. Fenton and Griffiths 2007; Liu et al. 2019b).

6.5.2.1 Spectral Representation Method

The SRM generates a random number as the superposition of a prescribed number of Fourier harmonics with random phase angle and possibly random amplitude. Suppose we want to generate a series of random numbers $X_n(t)$ of a variable t, with mean of 0, variance S_d^2 and auto-covariance function given by $S_d^2\rho(\tau)$, where $\rho(\tau)$ is the autocorrelation function for a given lag τ. The one-sided power spectral density function $S_x(\omega)$ can be readily derived from the auto-covariance function through the Wiener–Khintchine relations. For practical purposes, $S_x(\omega)$ is often treated as band-limited to frequency between 0 and $\bar{\omega}$, where $\bar{\omega} > 0$. For frequency less than 0 or higher than $\bar{\omega}$, $S_x(\omega) = 0$. If the frequencies are discrete, that is $\omega = \omega_1, \omega_2. . . .\omega_n$, then the variances S_{di}^2 for all the frequencies can be expressed approximately as (Grigoriu 1993)

$$S_{di}^2 \simeq S_x(\omega_i)\Delta\omega \tag{6.36}$$

where $\Delta\omega = \frac{\bar{\omega}}{n}$ and is known as the frequency discretisation. Equation (6.36) is approximately correct if $\frac{\bar{\omega}}{n}$ is small compared to $\bar{\omega}$, that is if n is large.

A common approach to generating the random quantity $X_n(t)$ is by the relation

$$X_n(t) = \left\{\sum_{i=1}^{n}\left[A_i S_{di}\cos(\omega_i t + \kappa_i)\right]\right\} \tag{6.37}$$

where κ_i are random phase angles uniformly distributed in [0, 2π] and A_i are Fourier amplitudes, which may be deterministic or random. If A_i are prescribed as deterministic variables, then the same value is typically adopted for all i's, that is $A_1 = A_2 =. . . A_n$. Grigoriu (1993) showed that both methods are approximately ergodic and Gaussian, provided the frequency discretisation is sufficiently fine; that is n is sufficiently large and the size of the random domain is large enough to provide adequate averaging of the random values. The SRM is usually implemented using fast Fourier transform to improve computational efficiency. The SRM for 2- and 3-D random fields are similarly defined by expanding Equation (6.37) into double and triple summations, respectively (details in Shinozuka and Deodatis 1996).

6.5.2.2 Turning Bands Method

The TBM is a method of simulating random two- and three-dimensional random fields based on pre-generated one-dimensional random fields along prescribed lines in the two- and three-dimensional spaces (e.g. Journel and Huijbregts 1978). This process is illustrated in Figure 6.5. A series of randomly oriented lines, RL_1, RL_2, . . . RL_L, are first generated through the space. Random values X_{ij} with zero-mean are then generated on discrete points along each of these lines, where the subscripts i and j denote the jth random value along RL_i. These random values can be generated by various points including SRM. To generate a random value at a point in the two- or three-dimensional spaces, normals are projected from the point to all the random lines. The pre-generated random values at the points of intersection between the respective normals and the random lines are then superimposed using the relationship

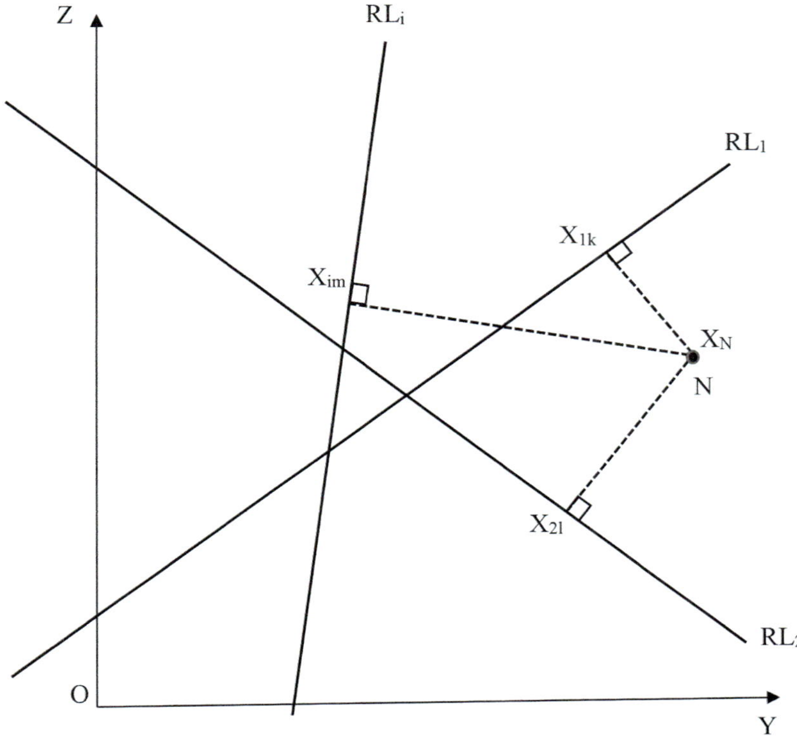

FIGURE 6.5 Schematic of the turning bands method in two-dimensional space. Lines RL_1, RL_2, . . . RL_n are randomly oriented lines. N represents the point for which the random value X_N is to be generated. X_{ij} denotes the jth random value on RL_i.

$$X_N = \frac{1}{\sqrt{L}} \sum_{i=1}^{L} X_{iJ} \tag{6.38}$$

where L is the number of randomly oriented lines and the subscript J denotes the random value at the intersection with the ith random line. The resulting random field is wide-sense stationary and isotropic and has covariance which is $\frac{1}{L}$ times that of the sum of the covariances of the random lines, resolved along respective lines.

6.5.2.3 Linear Estimation Method and Modified Linear Estimation Method

In the Linear Estimation method (Liu et al. 1986), the field of interest is first divided into n-noded finite elements of uniform size. These finite elements are prescribed purely for the purpose of generating the random field; they are different from the finite elements used in the stress analysis. The random variable $X(z)$ of any point within an element is calculated via the relation

$$X(z) = \sum_{i=1}^{n} N_i(z) \cdot X_i \tag{6.39}$$

where z is the position vector of a point within the element; X_i are given random variables at the ith nodal point and $N_i(z)$ are the corresponding shape functions of the finite element (Zienkiewicz and Taylor 2000). The shape functions satisfy the condition

$$\sum_{i=1}^{n} N_i(z) = 1 \tag{6.40}$$

Equation (6.40) is essentially an interpolation of the random numbers (i.e. X_i) at nodal points, which allows the value of the random variable to be generated at any prescribed location. The random numbers X_i are correlated and can be generated by various approaches including SRM. In this regard, the Linear Estimation method is similar to the TBM. To eliminate its dependency on other approaches (e.g. SRM), Liu et al. (2014) modified Equation (6.39) as

$$X(y) = \sum_{i=1}^{n} \sqrt{N_i(Jy + \varepsilon)} \cdot X_i \tag{6.41}$$

where J is a Jacobian matrix with a random rotation angle; ε is a translation vector with random components uniformly distributed in [0, 1]. The random numbers X_i are independent and identically distributed (i.i.d.) Gaussian numbers with zero mean and unit variance. As such, the modified method is an independent approach as it requires no other existing random field generation method (e.g. SRM). Liu et al. (2014) also verified the Gaussianity, stationarity and ergodicity of the random field generated by Equation (6.41).

6.6 PERFORMANCE OF SPATIALLY VARIABLE CHEMICALLY ADMIXED SOILS: RFE STUDIES

6.6.1 Axial Capacity of Cement-Treated Soil Columns

As discussed in sub-section 6.2.2, the axial capacity of treated soil columns in soft soil is dependent upon the unconfined compressive strength q_u of the treated soil. Equation (6.10) can also be expressed as

$$Q_{ult,col} = A_{col}\left(q_u + K_{p,col}\sigma_h\right) \tag{6.42}$$

so that, in absence of confining stress, $Q_{ult,col} = A_{col} q_u$. Equation (6.42) assumes that the unconfined compressive strength q_u is uniform throughout the treated soil column. In reality, there is significant spatial variation within the treated soil column. Hence, q_u is a point, rather than columnar, value. An approach that is often adopted in design is to prescribe an equivalent unconfined compressive strength q_{ueq} to account for the effect of spatial variation, that is

$$Q_{ult,col} = A_{col}\left(q_{ueq} + K_{p,col}\sigma_h\right) \tag{6.43}$$

An important objective of RFE analyses is to determine an appropriate value of q_{ueq} in relation to the distribution of q_u. In the Eurocode EC7 (BS EN 1997–1 2004), this equivalent value to account for spatial variation is termed the "characteristic value."

Namikawa and Koseki (2013) investigated the effects of spatial variation on axial capacity of treated soil columns, using RFE analyses and field data from Building Center of Japan (BCJ 1997). The field data from BCJ (1997) show a large spread in the unconfined compression capacity of the column, from approximately 0.3 to 1.15 times the mean strength of the core samples.

Namikawa and Koseki's (2013) RFE analyses were conducted using a randomised version of Namikawa and Mihira's (2007) cement-treated sand model. As discussed in Chapter 4, this model comprises linear yield and failure envelopes for shear and tensile loading, which describe yielding and failure at low and negative mean effective stresses. The model has no compression cap for yielding under isotropic and low $\left(q/p' \right)$ stress ratios. As discussed in Chapter 4, cement-treated sand usually does not undergo large volumetric compression.

The treated soil column analysed by Namikawa and Koseki (2013) has diameter of 1m and height of 2m. The analyses were assumed to be drained, which is reasonable as cement-treated sand is usually unable to retain excess pore pressure for long periods. The random unconfined compressive strength of treated soil was assumed to follow a truncated normal distribution with non-zero probability density between ±3.1 times the standard deviation of the mean. Based on their results, Namikawa and Koseki (2013) proposed that 50 realisations are sufficient to reflect the effect of spatial variation. Their results indicate that the autocorrelation distance Θ [Equation (5.6)] has a significant influence on the axial column capacity, with the mean column capacity reaching a minimum value at an autocorrelation distance of ~0.5m. Since the autocorrelation distance is half of the SoF, this implies that an SoF of about 1m, that is 1 column diameter, gives the lowest mean column capacity. Namikawa and Koseki's (2013) results also showed that, for a given CV of the element strength, all of the realisations give very similar column capacity when the autocorrelation length is zero. This is readily explained by the averaging effect of the column.

Pan et al. (2018a) also studied the effects of spatial variation on the capacity of cement-treated clay columns, using a randomised version of Xiao et al.'s (2017) cohesive Cam Clay model. This model features a yield locus with a shape that depends upon the cohesion, with the latter decreasing as cementitious bond in the improved soil is degraded, after yielding. When the cohesion is large, the yield locus has the shape of a steep arch. As the bonding is degraded, cohesion decreases to zero and the yield locus changes progressively to an elliptical shape, which is that of the Cam Clay model. This replicates the observed behaviour of cement-treated clay (Xiao et al. 2014).

Pan et al. (2018a) used 200 realisations in each random analysis. Three unconfined compressive strength distributions were studied, namely normal, log-normal and β-distributions. Pan et al.'s (2018a) study adopted the same column diameter as that of Namikawa and Koseki's (2013), that is 1m. In addition, they also investigated the effect of column length, drained versus undrained conditions as well as lateral confining pressure on column capacity. In addition, they also compared results obtained using the cohesive Cam Clay model with those obtained using the Mohr-Coulomb model.

Pan et al. (2018a) reported that failure generally occurred by shear banding due to presence of inhomogeneity. This is particularly so for the cohesive Cam Clay model under undrained loading, owing to post-yield strain softening arising from loss of bonding and decrease in mean effective stress. The decrease in effective stress is induced by volumetric collapse, which generates positive excess pore pressure under undrained condition. In the following discussion, the axial column capacity Q is defined as the peak failure load normalised by the cross-sectional area of the column. If soil strength is uniform within the column, then $Q = q_u$.

Pan et al.'s (2018a) results indicate that the SoF of unconfined compressive strength q_u has a significant influence on column axial capacity. When the SoF of q_u is much smaller than the column diameter, the mean column capacity approaches the mean unconfined compressive strength while the CV of column capacity, CV_Q, approaches zero. This finding is similar to that of Namikawa and Koseki's (2013) and indicates that there is a strong averaging effect within the column. As the SoF of q_u increases, so do the sizes of the weak and strong zones; and the averaging effect of the column size becomes less effective in minimising fluctuations in column capacity. This leads to an increase in CV_Q. In the extreme case, when the CV is much larger than the column diameter, each realisation, in effect, represents a uniform column, but with stochastically variable strength between different realisations. The mean column capacity also converges towards the mean q_u, but CV_Q approaches that of the unconfined compressive strength q_u, that is CV_q.

The lowest mean axial column capacity occurs when the SoF of q_u is approximately equal to the column diameter, which is also similar to Namikawa and Koseki's finding. In such a situation, the inherent weak zones tend to stretch across the entire column cross-section, enabling failure to occur at lower loads. Liu et al. (2015) also obtained a similar finding using a strain-softening Tresca model, in which the unconfined compressive strength decreases with strain post-peak. Their results also show a minimum column capacity being reached with the SoF of q_u is equal to the column diameter. Hence, the effect of SoF of q_u is largely independent of the constitutive model used. From a practical perspective, since each column is mixed and installed separately, the SoF induced by imperfect mixing cannot be larger than a column diameter. SoF larger than a column diameter is more likely to be due to in situ soil spatial variation rather than imperfect mixing.

Cases 5 and 6 in Pan et al.'s (2018a) study were analysed using the β-distribution with mean strength of 1.7MPa, CV of 0.4 and shape parameters α and β of 2.1 and 9.7, respectively. These parameters were fitted using the MBFC data discussed in Chapter 5. Table 6.2 summarises the results of these two cases. If the treated soil properties are uniform within the column, then $Q = q$. Subscripts $_A$ and $_5$ represent the average and 5th percentile values, respectively. E denotes the Young's modulus of the initial linear portion of the stress–strain curve. The first subscript following E, either c or s, refers to the overall column and spot modulus. The overall column modulus is defined as the modulus obtained from the average stress across the cross-section and the overall strain obtained from the compression of the ends of the column. The second subscript denotes the percentile. For instance, E_{cA} and E_{c5}

refer to the mean and 5th percentile overall modulus of the column. CV denotes the coefficient of variation of the parameter specified in the subscript. For example, CV_Q and CV_{Ec} denote the coefficient of variation of the peak axial capacity and overall modulus of the column. In Table 6.2, the column capacities are normalised by either the mean or the 5th percentile unconfined compressive strength while the overall modulus is normalised by the mean spot modulus.

TABLE 6.2

Summary of some results from Cases 5 and 6 of Pan et al.'s (2018a) RFE analysis, using β-distribution.

		Effective confining pressure/q_{uA}			
Drainage condition	**Items**	**0**	**0.035**	**0.07**	**0.14**
Undrained	Q_A/q_{uA}	0.67	0.68	0.70	0.70
	CV_Q	0.26	0.24	0.24	0.24
5th percentile strength reduction ratio	Q_5/q_{uA}	0.39	0.42	0.42	0.42
	α_{Q5}	1.53	1.45	1.45	1.45
	Q_A/q_{u5}	1.41	1.43	1.47	1.47
	Q_5/q_{u5}	**0.82**	**0.88**	**0.88**	**0.88**
Estimated Q_{p5} based on normal distribution in column capacity	Q_5/q_{uA}	0.38	0.41	0.42	0.42
	Q_5/q_{u5}	0.81	0.86	0.89	0.89
	E_{cA}/E_{sA}	0.82	0.87	0.90	0.89
	CV_{Ec}	0.20	0.15	0.13	0.10
	E_{c5}/E_{sA}	0.55	0.65	0.70	0.74
Drained	Q_A/q_{uA}	0.63	0.61	0.55	0.52
	CV_Q	0.26	0.26	0.26	0.33
Strength reduction ratio	Q_5/q_{uA}	0.36	0.34	0.32	0.24
	α_{Q5}	1.60	1.65	1.70	1.90
	Q_A/q_{u5}	1.32	1.28	1.15	1.09
	Q_5/q_{u5}	**0.76**	**0.71**	**0.67**	**0.50**
Estimated Q_{p5} based on normal distribution	Q_5/q_{uA}	0.36	0.35	0.31	0.24
	Q_5/q_{u5}	0.76	0.73	0.66	0.50
	E_{cA}/E_{sA}	0.95	0.91	0.87	0.83
	CV_{Ec}	0.24	0.18	0.15	0.15
	E_{c5}/E_{sA}	0.57	0.64	0.66	0.62

Q and q_u represent column capacity and unconfined compressive strength, respectively. Subscripts A and 5 represent the average and 5th percentile values, respectively. Estimated Q_5 values are based on the assumption that the axial column capacity is normally distributed. E denotes the Young's modulus of the initial linear portion of the stress–strain curve. The first subscript following E, either $_c$ or $_s$, refers to the overall column modulus and point modulus, respectively.

Data source: Pan et al. (2018a).

Pan et al.'s (2018a) results indicated that, although a positively skewed β-distribution was used to describe the spatial variation in spot strength, the computed column capacity follows an approximately normal distribution. As Table 6.2 shows, the 5th percentile column capacity Q_5, estimated from Q_A and CV_Q, assuming a normal distribution, agrees well with that evaluated from the realisations. However, if the β-distribution is assumed for unconfined compressive strength q_u, then q_u is bounded, and the column capacity is also likely to be bounded. An absolute upper bound of the peak column capacity Q_p would be reached when the column is uniform and has a (peak) unconfined compressive strength q_u equal to the upper bound q_u since all points will reach the peak strength simultaneously. The lower bound Q_p is more uncertain since different points within a non-uniform column may reach the peak q_u at different instances. Hence, the lower bound Q_p may be influenced by the degree of post-peak strain softening and may not be equal to that of a uniform column with lower bound (peak) q_u. Nonetheless, there will still be a positive lower bound equal to the critical state (ultimate) q_u of a uniform column with lower bound q_u. Hence, the distribution of the column capacity Q_p is likely to be bounded. Since the normal distribution is unbounded, extrapolation of Q_p towards the tail ends, beyond the data, using the normal distribution is likely to be significantly in error. A truncated normal distribution or other bounded distributions (e.g. β-distribution) may overcome this disadvantage of the normal distribution.

Pan et al. (2018a) also investigated the effect of the length-to-diameter (*L/D*) aspect ratio of the column. As the aspect ratio increases, the mean and CV of column capacity decrease. This is because the chances of a localised shear plane encountering weak zones with very low unconfined compressive strength in each realisation increases with aspect ratio.

6.6.2 Lateral Capacity of Cement-Treated Soil Slabs for Deep Excavation Construction

In deep excavation projects in very soft clay conditions, excessive ground movement, settlement and retaining wall instability are often a major concern. This is particularly so in urban built-up environments, wherein excavations often take place in close proximity to buildings and structures. In such cases, excessive wall movement will lead to excessive ground movement, settlement and even collapse, jeopardising neighbouring structures. A well-known example of excavation collapse in soft clay is the Nicoll Highway collapse during the construction of the Circle Line in Singapore (Ministry of Manpower, Singapore 2005).

Above excavation level, struts can be used to control wall deflection and ground movement. However, in soft ground conditions, the maximum wall deflection is often reached below final formation level, where it is infeasible to use struts. In such situations, cement-treated soil layers, termed hereafter as "cement-treated soil slabs," are often used to function as underground struts to control wall deflection (e.g. Nakagawa et al. 1996; Lee et al. 1998; O' Rourke et al. 1998; Shirlaw et al. 2000a, 2000b, 2006).

In current design, cement-treated soil slabs are often assumed to be isotropic and homogeneous. In reality, the improved soil is significantly non-uniform owing to imperfect mixing (e.g. Liu et al. 2017) as well as spatial variation in in situ soil conditions. In addition, it has a strongly columnar structure (e.g. Nakagawa et al. 1996) due to the column-by-column mixing process. Data from deep-mixing and jet-grouting projects in Singapore indicate that the coefficient of variation in unconfined compressive strength derived from core sample typically ranges from 0.39 to 0.46. Cement-treated soil slabs are typically designed as isotropic and homogeneous structures, but with an equivalent strength, termed herein as "mass strength," which is assumed to be several times lower than the strength from laboratory-moulded samples with the same mix proportion.

Liu et al. (2015) investigated the undrained behaviour of cement-treated soil slab using a total stress Tresca model with undrained cohesion taken to be half of the unconfined compressive strength q_u. Post-peak strain softening was modelled by reducing the undrained cohesion to 0.8 times its peak value when a threshold shear strain of 1.5% was exceeded. The modulus was assumed to be a constant multiple of the undrained cohesion.

Three types of spatial variation were modelled. The first is a deterministic radial variation. Three trends of deterministic radial variation were studied: constant strength across the radial section, as well as linearly increasing and decreasing strength with radius. The second type of spatial variation is intra-column stochastic variation in strength, this being characterised by a mean strength, as well as its CV and SoF. Liu et al. (2015) used a β-distribution with shape parameters $\alpha = 1.86$ and $\beta = 3.78$. The CV is varied as part of the parametric studies while the mean unconfined compressive strength was kept constant at approximately 2.15 MPa. As results are normalised by the mean unconfined compressive strength, its specific value is not pertinent. The overall modulus of the slab, defined as the average stress across the ends of the slab divided by the overall strain, is also normalised by the average point modulus.

The third type of spatial variation is that arising from positioning error of the cement-treated soil column. Such positioning error can be caused by positioning error of the mixing/grouting rig and off-verticality of the mixing/grouting stem (Croce and Modoni 2007). Table 3.4 shows some reported permissible horizontal offset from prescribed position and off-verticality. In general, the permissible off-verticality is between 1% (~0.6°) and 2% (~1.1°). For shallow improvement works, the horizontal offset may contribute more to column positioning error. However, for deep improvement works, the off-verticality often contributes more. To illustrate, if the improvement soil layer lies at a depth of 20m below ground surface, a 1% off-verticality will result in a column offset of 200mm at the improved depth; this is much larger than error from horizontal offset. In Liu et al. (2015), positioning error was modelled by prescribing a lateral offset around the column centre. This lateral offset follows a two-dimensional random field with uniform marginal distribution, both radially and azimuthally, subjected to a maximum deviation, which is used as a characterising parameter in their study.

Liu et al. (2015) defined the mass strength of the slab as the maximum lateral compressive force which can be sustained by the slab divided by the slab cross-sectional area. Their results showed that the mass strength follows an approximately normal distribution, even though the unconfined compressive strength follows a

β-distribution. The CV of the mass strength is also lower than that of the unconfined compressive strength, being generally less than 0.1. This suggests a significant averaging effect.

The effects of the deterministic radial trend on the mass strength and overall modulus of the slab are also relatively insignificant when the quantities are normalised by their respective volume-averaged values. The radial, vertical and circumferential SoFs were also found to have insignificant effect if they are of the order of one column diameter. This differs markedly from the effect of SoF on the axial capacity of treated soil columns and can be readily explained in terms of the SoF relative to the size of the geotechnical structure. When non-contiguous treated soil columns are loaded axially, each column behaves as a geotechnical structure. Hence, the significant dimension of this geotechnical structure is its diameter; an SoF that is equal to the column diameter is also equal to the significant structural dimension. In such situations, the typical size of the weak zone is equal to the size of the geotechnical structure and can dominate the behaviour of the structure. On the other hand, a cement-treated soil slab typically comprises numerous treated soil columns and has a much larger dimension. Its significant dimension is either the length or the width of the slab. An SoF that is equal to a column diameter implies that the size of weak zones is still much smaller than the significant dimension of the slab. Hence, significant averaging effect can still be obtained.

Liu et al. (2019a) noted that the SoF structure of deep-mixed soil may be dual ranged. The SoF arising from imperfect mixing is likely to be short ranged. Circumferential SoF cannot exceed 360° or one column circumference. Moreover, since every column is cast separately, imperfect mixing is unlikely to result in radial SoF greater than one diameter. Although SoFs much larger than a column diameter have been reported (e.g. Filz and Navin 2006), these are probably due to other factors such as variation in in situ soil conditions. This type of long-ranged spatial variation was not modelled by Liu et al. (2015). However, the averaging effect on such long-ranged spatial variation is likely to be weaker. In such situations, the influence of SoF may be more significant.

While the SoF characterises the size range of the weak and strong zones, the CV characterises the overall range of variation of the parameter. A large CV in the spot strength implies a highly non-uniform distribution in strength within a column, with many locations having very low and high strength. Conversely, a small CV implies a relatively uniform strength distribution. Liu et al. (2015) noted that, if the CV of intra-column stochastic fluctuation in point strength lies below about 0.4, the mass compressive strength of the slab is not significantly impacted. As the intra-column CV rises above 0.4, the mass compressive strength of the slab starts to decrease at an increasing rate with CV. This trend is illustrated in Figure 6.6. A similar trend exists for the effect of point modulus on overall modulus, that is the modulus of the entire slab.

The normalised positioning error, defined as the maximum lateral deviation ξ from the prescribed position normalised by the column radius R, also has a significant effect on the mass strength and overall modulus of the slab. If the normalised positioning error ξ/R exceeds about 0.35, the average as well as 5th percentile mass strength starts to decrease at a much faster rate (Figure 6.7). Physically, the effect of positioning error is to leave a volume of untreated soil with very low strength. Furthermore, positioning error will lead to an increase in the CV of unconfined compressive strength in the slab, over and above that arising from intra-column stochastic variation. Hence, as illustrated in Figure 6.6, the CV in unconfined compressive strength that one would obtain by random sampling of the slab, termed herein as "resultant CV", is higher than that due to intra-column stochastic variation alone.

Liu et al. (2015) also proposed relationships for estimating the equivalent mass strength and overall stiffness of the cement-treated soil slab from the mean and CV in unconfined compressive strength of the slab, of the form

$$Q_d = \left[\frac{Q_A\left(1-\beta CV_Q\right)}{1.1}\right] \tag{6.44}$$

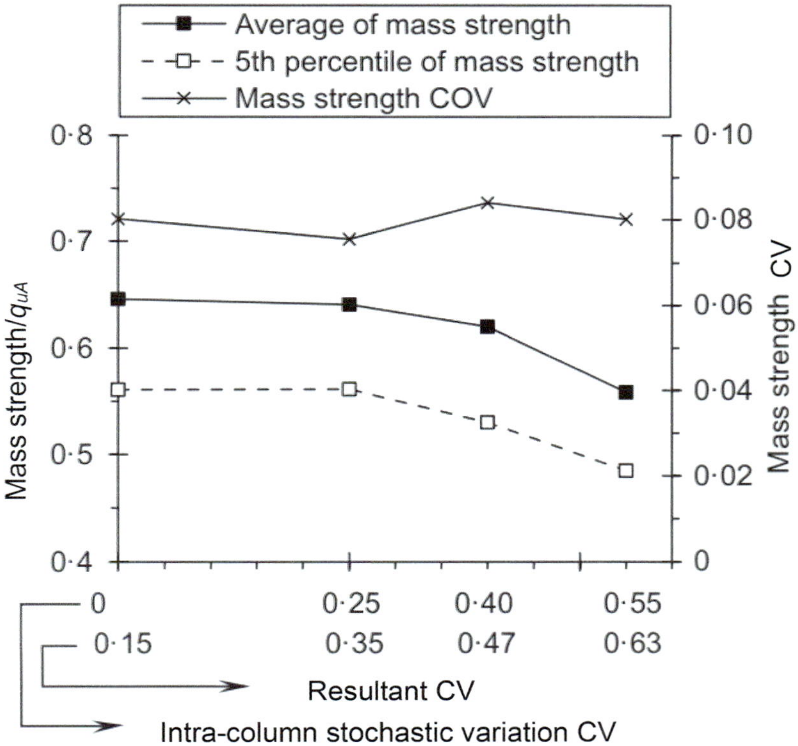

FIGURE 6.6 Effect of intra-column and resultant CV on the mean and CV of mass strength. Deterministic radial strength variation: inner stiffer with column centre strength twice the peripheral strength. Positioning error: 50% of radius.

After Liu et al. (2015).

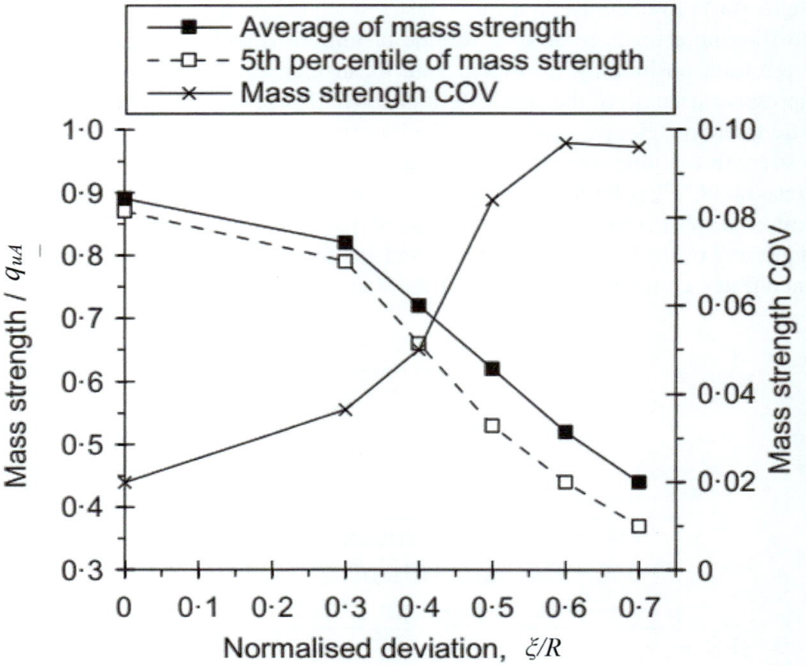

FIGURE 6.7 Effect of positioning error on the mean and CV of mass strength. Deterministic radial strength variation: inner stiffer with column centre strength twice the peripheral strength. Resultant CV: 0.47.

After Liu et al. (2015).

where Q_{d} is the design strength corresponding to the reliability index β, Q_A is the mean mass strength and $\mathrm{CV_Q}$ is the CV of the mass strength. Liu et al. (2015) proposed the factor of 1.1 in Equation (6.45) to account for potential bias arising from the limited number of simulations. More details are available in Liu et al. (2015). Equation (6.44) can also be written in the form

$$Q_d = q_{\mathrm{uA}}\left[1-\alpha_q \mathrm{CV}_q\right] \tag{6.45}$$

$$\alpha_q = \frac{1-\dfrac{Q_{\mathrm{A}}\left(1-\beta \mathrm{CV_Q}\right)}{1.1 q_{\mathrm{uA}}}}{\mathrm{CV_{qr}}} \tag{6.46}$$

where q_{uA} is the mean unconfined compressive strength and $\mathrm{CV_q}$ and $\mathrm{CV_{qr}}$ are the CV of the unconfined compressive strength due to stochastic variation alone and resultant CV, respectively, that is the CV of unconfined compressive strength considering stochastic variation and positioning error. As the inset of Figure 6.8 illustrates, α_{Q} is a strength reduction factor, akin to the reliability index, except that the latter

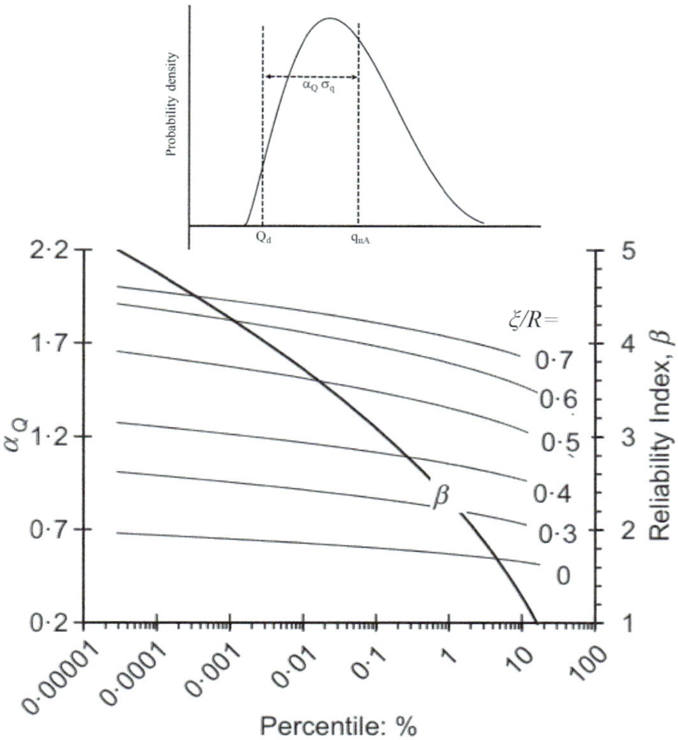

FIGURE 6.8 Values of α_Q and β against percentile of mass strength and positioning error η normalised by the column radius R for resultant UCS CV of 0.47. Inset illustrates statistical concept of α_Q.

After Liu et al. (2015).

applies to the factor or margin of safety instead of the strength. Liu et al. (2015) also proposed a similar set of relationships for the overall modulus of the slab.

Pan et al. (2018b) also studied the undrained and drained performance of cement-treated soil slab using the β-distribution, but with an effective stress cohesive Cam Clay model (Xiao et al. 2017). Pan et al.'s (2018b) load–displacement curves all showed peak mass strength with post-peak softening. This is different from Liu et al.'s (2015) results, which show the load rising to a stable ultimate value. This arises because the cohesive Cam Clay model has much more pronounced peak strength and rapid post-peak softening than Liu et al.'s (2015) Tresca model.

Pan et al.'s (2018b) undrained random analysis results shows very similar trends as Liu et al.'s (2015), except that the mean and 5th percentile mass strength is consistently higher by about 10%. This is because Pan et al. (2018b) used the peak mass strength whereas Liu et al. (2015) used the ultimate strength. Under drained condition, however, the behaviour of the slab is heavily influenced by volumetric collapse of the treated soil. The load–displacement curves indicate yielding initiating at a relatively low load, followed by strain-hardening, which is similar to that reported by Xiao et al. (2017) for triaxial tests. This is due to decrease in void ratio arising from volumetric reduction, which increases the strength of the treated soil and counteracts the softening due to loss of bonding. Critical state is reached at a very large strain exceeding 20%. For this reason, Pan et al. (2018b) recommended the use of a "yield stress" of the slab, which they defined to be the turning point between the initial linear region and the subsequent strain-hardening region. Table 6.3 summarises some of Liu et al.'s (2015) and Pan et al.'s (2019) results.

Pan et al. (2018b) noted that, apart from the CV of unconfined compressive strength and positioning error, the effective confining pressure p' also influences the mass strength, overall modulus and their coefficients of variation, especially under drained condition. This effect cannot be reflected in Liu et al.'s (2015) total stress model. Pan et al. (2018b) proposed a modification of Liu et al.'s (2015) design approach, by first expressing the design strength Q_d in terms of the mean mass

TABLE 6.3

Summary of some results from Liu et al. (2015) and Pan et al. (2019) on underground cement-treated soil slabs.

References		**Liu et al.'s (2015) results**		**Pan et al.'s (2018) results**	
Drainage condition and names of parameters		**Normalised positioning error (ξ/R)**			
Undrained	**Parameters**	**0.5**	**0.7**	**0.5**	**0.7**
Mean equivalent strength/mean UCS	Q_A/q_{uA}	0.62	0.44	0.64	0.42
CV of equivalent strength	CV_Q	0.08	0.10	0.09	0.14
5th percentile strength reduction ratio	Q_5/q_{uA}	0.53	0.38	0.54	0.35
5th percentile strength reduction factor	α_{Q5}	1.00	1.32	0.98	1.38
Mean equivalent strength/5th percentile UCS	Q_A/q_{u5}	1.88	1.34	1.94	1.28
Ratios of 5th percentile equivalent strength to various percentiles UCS	Q_5/q_{u5}	1.61	1.15	1.64	1.06
	Q_5/q_{u20}	0.94	0.67	0.96	0.62
	Q_5/q_{u10}	1.26	0.91	1.29	0.83
	Drained				
Mean equivalent strength/mean UCS	Q_A/q_{uA}			0.37	0.25
CV of equivalent strength	CV_Q			0.075	0.163
5th percentile strength reduction ratio	Q_5/q_{uA}			0.34	0.2
5th percentile strength reduction factor	α_{Q5}			1.40	1.70
Mean equivalent strength/5th percentile UCS	Q_A/q_{u5}			1.12	0.76
5th percentile equivalent strength/5th percentile UCS	Q_5/q_{u5}			1.03	0.61

Data sources: Liu et al. (2015), Pan et al. (2019).

strength Q_A and CV of mass strength CV_Q as well as the reliability index β, via the relationship

$$Q_d = Q_A\left(1 - \beta CV_Q\right) \tag{6.47}$$

Q_A is then related to mean unconfined compressive strength q_{uA} using the relationship

$$Q_A = R_1 R_2 q_{uA} \tag{6.48}$$

wherein the factors R_1 and R_2 are reduction factors for positioning error and effective confining pressure, respectively. As Figures 6.9a and b show, positioning error decreases mean strength and overall modulus under both drained and undrained conditions. It should be noted that Pan et al. (2018b) defined the normalised positioning error as maximum deviation from intended column position divided by the column diameter, rather than column radius. The mass strength and overall modulus also decrease with effective confining pressure, especially under drained condition, wherein an increase in the effective confining pressure predisposes the weak zones towards volumetric collapse, thereby decreasing the overall modulus and yield strength.

Pan et al. (2018b) also proposed that the CV_Q be expressed in terms of coefficient of variation of mass strength at zero effective confining pressure (CV_{Q0}) via the relationship

$$CV_Q = R_4\left(CV_{Q0}\right) \tag{6.49}$$

where R_4 is a modification factor which accounts for the effect of mean effective stress. As Figure 6.10a shows, CV_{Q0} increases with positioning error, this trend being

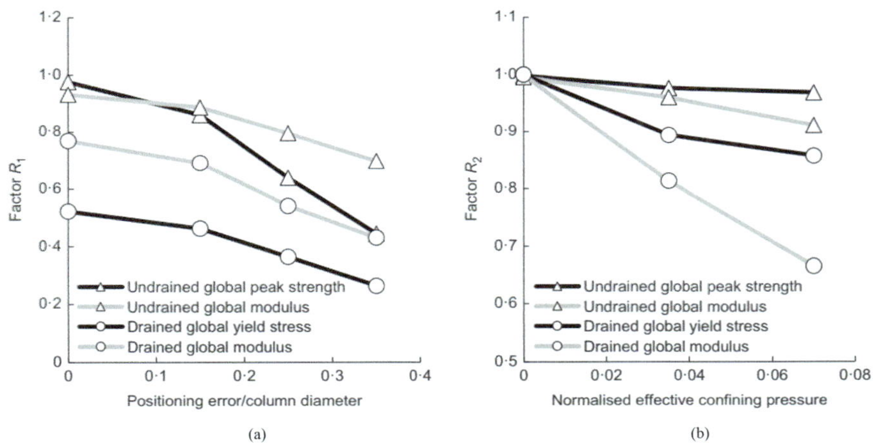

FIGURE 6.9 Effect of (a) positioning error and (b) effective confining pressure, respectively, on factors R_1 and R_2, which influence the mean mass strength of slab.

After Pan et al. (2018b).

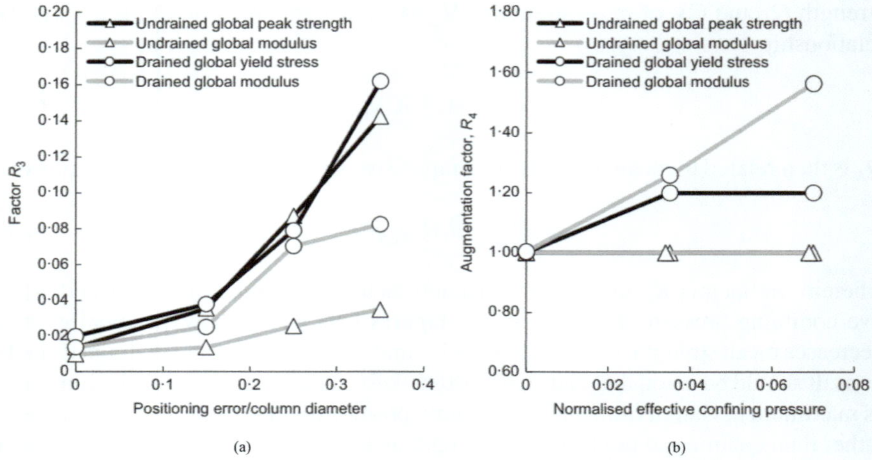

FIGURE 6.10 (a) Effect of positioning error on CV_{Q0} and (b) effect of effective confining pressure on R_4.

After Pan et al. (2018b).

similar to that of Liu et al. (2015). Increasing the effective confining pressure also increases the CV_Q under drained condition, this being reflected in the augmentation factor R_4 in Figure 6.10b. Under undrained condition, the effective confining pressure has no effect on CV_Q and therefore slab performance. On the other hand, under drained condition, increasing the effective confining pressure degrades the mass strength and overall modulus. This causes a more significant decrease in mass strength and an increase in the CV of mass strength and overall modulus.

6.6.3 Tunnel Construction Through Cement-Treated Soil

Chemical soil improvement has been used for support in soft ground tunnelling. One mode of application is to use contiguous jet-grouted or deep-mixed columns to form an improved soil surround around the extrados of the tunnel or an "umbrella" around the tunnel roof to reduce ground movement, enhance face stability and facilitate tunnelling without compressed air. The effect of spatial variability in improved soil properties on the performance of the improved soil surround has been studied by Tyagi et al. (2018), Pan and Lee (2019), Pan et al. (2020) and Tyagi et al. (2020), assuming idealised improved soil surrounds of uniform thickness around circular tunnels.

As Table 6.4 shows, three of the studies used two-dimensional plane strain RFE analyses, which assume that the distribution of properties, for example strength, is uniform in the out-of-plane direction. This is equivalent to having an SoF of infinity in the out-of-plane direction. Ching et al. (2016) noted that the results from the two-dimensional plane-strain analysis are generally more pessimistic than those of corresponding three-dimensional analysis because weak zones are replicated over an infinitely long distance in the out-of-plane direction, and averaging effect cannot come into play in this direction. Tyagi et al. (2020) used three-dimensional analyses replicating columnar structure with different SoFs in the radial, azimuthal (circumferential) and vertical directions.

In all four studies, the tunnel models are supported by uniform pressure applied to the inner wall of the improved soil surround. The support pressure is progressively reduced until collapse or large displacement initiates; this critical support pressure is taken to represent the capacity of the improved soil surround. In addition to the RFE analyses, the studies also included deterministic analyses, assuming that the improved soil surrounds are homogeneous. By equating the critical support pressure of the random analyses to that of deterministic analyses, the equivalent strength Q_i of an equivalent homogeneous tunnel, which gives the same critical tunnel support pressure as the ith percentile from the random analysis results, can be determined. The ith percentile also quantifies the probability of failure. For instance, Q_5 is the equivalent strength of a homogeneous improved soil surround which has the same critical support pressure as the spatially variable soil surround with 5% probability of failure. This allows a normalised equivalent strength R_i to be defined by

$$R_i = \frac{Q_i}{q_{\mathrm{uA}}} \tag{6.50}$$

where q_{uA} is the mean unconfined compressive strength and the subscript i represents the ith percentile of the RFE support pressure to which the equivalent deterministic strength is equated. Using Equations (6.45) and (6.46), and substituting Q_i for Q_{d}, the ith percentile strength reduction factor α_{Qi} can also be written as

$$\alpha_{Qi} = \frac{1}{\mathrm{CV_q}}\left(1 - \frac{Q_i}{q_{\mathrm{uA}}}\right) \tag{6.51}$$

One of the parameters which have a significant influence on R_5 is $\mathrm{CV_q}$, which varies significantly even within the same study. In general, α_{Q5}, that is the value of strength reduction factor corresponding to 5% probability of failure, decreases as $\mathrm{CV_q}$ increases. This indicates that the decrease in equivalent strength is less than proportionate compared to $\mathrm{CV_q}$.

6.6.4 Other Studies

Tyagi and Tamta (2023) reported a random finite element studies of cement-treated soil slopes. For $\mathrm{CV_q}$ of 0.45, the α_{Q5} from their studies vary from 0.67 to 0.82 as the mean unconfined compressive strength varies from 0.5MPa to 4MPa. The ratio of horizontal-to-vertical SoF was found to have an insignificant effect. However, the ratio of horizontal SoF to slope width was found to have a significant effect.

Kek et al. (2021) examined the effect of spatial variation in binder content on one-dimensional consolidation behaviour of a layer of cement-treated subgrade, considering the influence of binder content variation on permeability as well as strength. The void ratio of the cement-treated soil was first determined using Tyagi et al.'s (2019) approach, which involved estimating the amount of cementitious material produced.

TABLE 6.4

Summary of random finite element studies on tunnel improved soil surrounds.

	Geometry				Input parameters				Cement-treated soil UCS				Results for 5% probability of failure	
Reference	Two- or three-dimensional	Tunnel diameter D (m)	Thickness-to-diameter ratio of improved soil surround (t/D)	Soil cover-to-diameter ratio (C/D)	Total/ effective stress	Drainage conditions	Material model	Spot strength distribution	Mean strength (kPa)	CV of UCS (CV_q)	SoF-to-thickness ratio (SoF/t)	Positioning error/ column radius ξ/R	Normalised 5th percentile strength of equivalent homogeneous tunnel $R_5 = Q_5/q_{uA}$	$\alpha_{Q5} = (1 - Q_i/q_{uA})/CV_q$
Tyagi et al. (2018)	2D	15	0.2	1.33	Effective stress	Undrained	Cement-treated soil: Mohr-Coulomb Soft soil: Cam Clay	Log-normal	2800	0.2, 0.3, 0.4, 0.5, 0.55, 0.6	0.16–0.67	Not applicable	Range: 0.3–0.48. For CV_q of 0.4, $R_5 = 0.4$.	For CV_q of 0.2 to 0.6: $\alpha_{q5} =$ 2.6–1.16. For CV_q of 0.4, $\alpha_{q5} = 1.5$.
Pan and Lee (2019)	2D	11, 12, 15	0.2–0.45	1.0–1.64	Effective stress	Undrained	Cement-treated soil: Cohesive Cam Clay Soft soil: Cam Clay	log-normal	600, 1200, 1700	0.25, 0.4, 0.55	0.16–2.67	Not applicable	Range: 0.4–0.66. For CV_q of 0.4, $R_5 = 0.52$.	For CV_q of 0.25 to 0.55: $\alpha_{q5} =$ 1.36–1.1. For CV_q of 0.4, $\alpha_{q5} = 1.21$.

Tyagi et al. (2020)	3D with columnar structure	15	0.2	1.3	Effective stress	Consolidation. Dimensionless time T = 0.0002, 0.2 and 200	Cement-treated soil: Mohr-Coulomb Soft soil: Cam Clay	β-distribution	2800	0.4, 0.55*	Depth: 0.63–2.5 Radial: 0.042–0.17. Azimuthal: $\pi/8$–$\pi/2$ rads.	0, 0.3, 0.5, 0.7	Based on reference case parameters. Range for variable ξ/R: For CV_q of 0.4 and ξ/R = 0–0.7: 0.36–0.48. For CV_q of 0.55 and ξ/R = 0–0.7: 0.31–0.45.	5% probability of failure **based on reference case parameters**: (a) CV_q = 0.4 and p/r = 0–0.7: α_{Q5} = 1.3–1.6. (b) CV_q = 0.55 and p/r = 0–0.7: α_{Q5} = 1.0 to 1.25. Likely to be partially drained.
Pan et al. (2020)	2D	8, 12 (reference case), 20	0.15, 0.25 (reference case), 0.35	1.0, 1.42 (reference case), 2.0	Effective stress	Drained and undrained	Cement-treated soil: Cohesive Cam Clay/ Mohr-Coulomb/ Tresca Soft soil: Cam Clay	log-normal	600 (reference case), 1200, 1700	0.25, 0.4 (reference case), 0.55	1.0	Not applicable	For D = 12 m, t/D = 0·25, C/D = 1·4, q_a = 600 kPa, UCS CV_q = 0·4, SoF/t = 1·0, 5% probability of failure: (a) Tresca: drained and undrained 0.54, (b) Mohr-Coulomb: drained and undrained 0.46, (c) Cohesive Cam Clay: drained 0.45, undrained 0.49.	For D = 12 m, t/D = 0·25, C/D = 1·4, q_a = 600 kPa, UCS CV = 0·4, SoF/t = 1·0: (a) Tresca: drained and undrained 1.16, (b) Mohr-Coulomb: drained and undrained 1.35, (c) Cohesive Cam Clay: drained 1.37, undrained 1.28.

* Does not include the component of CV arising from positioning error.

The coefficient of permeability was then estimated using a relationship which was fitted using available experimental data. The unconfined compressive strength of the treated soil was estimated using Chen et al.'s (2016) relationship. This allows a relationship between the strength and coefficient of permeability to be deduced. The relationship depends significantly upon the water–cement ratio of the cement slurry. For water–cement slurry of approximately 1.0 and below, the coefficient of permeability decreases as the strength increases. This is because as the cement slurry content increases, the strength increases while the void ratio decreases as more cementitious material fills up the voids. On the other hand, for very high water–cement ratio of about 2 or higher, the coefficient of permeability increases as the strength increases. This is because as the cement slurry content increases, so does the water content and void ratio of the treated soil. However, this anomalous scenario is unlikely to arise in practice, as the typical water–cement ratio used is about 1.0 or slightly less. Hence, in general, the cement-treated soil should become more impermeable and stronger as the binder content increases. Kek et al. (2021) studied the problem of a layer of cement-treated soil consolidating under vertical loading. In a uniform scenario, this is the classical one-dimensional consolidation problem. Their findings indicated that where the horizontal SoF is much larger than the vertical SoF, the stronger, less permeable zones and the weaker, more permeable zones are horizontally layered. In such situations, yielding of the weaker zones will increase the positive excess pore pressure in these zones. At the same time, the stronger, less permeable zones retard pore pressure dissipation, thereby reducing the rate of consolidation. On the other hand, where the vertical SoF is much longer than the horizontal SoF, the stronger, less permeable zones and the weaker, more permeable zones are vertically panelled. This allows the stronger zones to attract higher loading and thereby generate higher positive excess pore pressure, while the weaker and more permeable zones serve as more permeable vertical drainage pathways, enabling pore pressure dissipation, similar to vertical drains, and thus hastening the consolidation process. In a scenario where the SoFs in the horizontal and vertical directions are roughly similar, the consolidation rate will depend on whether load distribution or pore pressure dissipation screening has the greater effect. In general, the consolidation rate will lie between that of the vertically panelled and horizontally layered extremes.

6.7 PRACTICAL IMPLICATIONS

6.7.1 Design Implications

The spatially variable nature of soils has long been recognised. In the past, methods of accounting for spatial variation in design were often left to the judgement of engineers. The Eurocode EC7's (BS EN 1997–1 2004) stipulation of characteristic values marks an important step towards reducing the uncertainty that spatial variation can pose to engineers. The EC7 stipulates the following:

> (1) The selection of characteristic values for geotechnical parameters shall be based on results and derived values from laboratory and field tests, complemented by well-established experience.

(2) The characteristic value of a geotechnical parameter shall be selected as a cautious estimate of the value affecting the occurrence of the limit state.
(3) The greater variance of c' compared to that of $\tan\varphi'$ shall be considered when their characteristic values are determined.
(4) The selection of characteristic values for geotechnical parameters shall take account of the following:
- geological and other background information, such as data from previous projects;
- the variability of the measured property values and other relevant information, for example from existing knowledge;
- the extent of the field and laboratory investigation;
- the type and number of samples;
- the extent of the zone of ground governing the behaviour of the geotechnical structure at the limit state being considered;
- the ability of the geotechnical structure to transfer loads from weak to strong zones in the ground.

....

(7) The zone of ground governing the behaviour of a geotechnical structure at a limit state is usually much larger than a test sample or the zone of ground affected in an in situ test. Consequently the value of the governing parameter is often the mean of a range of values covering a large surface or volume of the ground. The characteristic value should be a cautious estimate of this mean value.
(8) If the behaviour of the geotechnical structure at the limit state considered is governed by the lowest or highest value of the ground property, the characteristic value should be a cautious estimate of the lowest or highest value occurring in the zone governing the behaviour.
(9) When selecting the zone of ground governing the behaviour of a geotechnical structure at a limit state, it should be considered that this limit state may depend on the behaviour of the supported structure. For instance, when considering a bearing resistance ultimate limit state for a building resting on several footings, the governing parameter should be the mean strength over each individual zone of ground under a footing, if the building is unable to resist a local failure. If, however, the building is stiff and strong enough, the governing parameter should be the mean of these mean values over the entire zone or part of the zone of ground under the building.
(10) If statistical methods are employed in the selection of characteristic values for ground properties, such methods should differentiate between local and regional sampling and should allow the use of a priori knowledge of comparable ground properties.

If statistical methods are used, the characteristic value should be derived such that the calculated probability of a worse value governing the occurrence of the limit state under consideration is not greater than 5%.
NOTE: In this respect, a cautious estimate of the mean value is a selection of the mean value of the limited set of geotechnical parameter values, with a confidence level of 95%; where local failure is concerned, a cautious estimate of the low value is a 5% fractile.

This excerpt from BS EN 1997–1 (2004) indicates that one function of the characteristic value is to account for variation in the measured property, such as strength. Many of the BS EN 1997–1 (2004) recommendations are general and leave substantial room for engineering judgement. However, there are explicit guidelines on the use of a cautious estimate of the mean value or individual measurements, the important criterion being the ability of the superstructure to redistribute loading should local failure occur in a weak zone. In the following discussion, the applicability of EC7's recommendations to treated soil structures will be examined.

6.7.1.1 Scale of Fluctuation and Averaging Effect

6.7.1.1.1 Axially Loaded Treated Soil Columns

The random finite element studies on spatial variation of chemically improved soils indicate that, in some scenarios, an important parameter to consider is the SoF of spatial variation of the treated soil mass. This is not considered by EC7 (BS EN 1997–1 2004). To illustrate, let us consider a hypothetical example of a footing supported by a cement-treated soil column. If the footing is sufficiently stiff, Clause 9 of EC7's recommendation would permit the use of the mean strength of the ground beneath the footing, which implies using the mean strength of the treated soil column. However, as Table 6.2 shows, under undrained situation, the mean column capacity Q_A is only about 67% to 70% of the mean unconfined compressive strength of the column. This is because the worst-case scenario occurs when the SoF is approximately one column diameter, at which point the area of the weak zones is similar to the column cross-sectional area. Under such conditions, the averaging effect is not fully functional.

As Table 5.4 shows, the radial SoF is typically of the order of 0.12 to 0.25 times the column diameter while the circumferential SoF is typically 1 to 2 radians. At the column periphery, this translates to a circumference of roughly 0.5 to 1 column diameter. Hence, although the radial and circumferential SoFs are not exactly one column diameter, they are, nonetheless, significant proportions of a column diameter. Since the SoF may vary, it is probably prudent to adopt the cautious worst-case assumption of one column diameter.

The EC7 also recommends that in situations where local failure can occur, such as the failure of an individual footing, then a 5th percentile strength should be used instead of a 95% confidence level mean value. As Table 6.2 shows, under undrained condition, the 5th percentile column capacity Q_5 is about 82% to 88% of the 5th percentile UCS q_{u5}, since the averaging effect is offset by the SoF effect. Since the column capacity is also the equivalent strength of a uniform column, Q_5 is the equivalent strength that is exceeded by 95% of the cases. Q_5, rather than q_{u5}, can therefore be surmised to be the value satisfying the EC7 requirement that "the characteristic value should be . . . such that the calculated probability of a worse value governing the occurrence of the limit state under consideration is not greater than 5%." Conversely, using q_{u5} would, under undrained condition, translate to an equivalent column strength Q that lies between the 9th and 13th percentiles of the equivalent strength distribution, this percentile being also the probability of failure. Under drained condition, the percentile is even higher, between the 17th and 29th percentiles. This is significantly higher than the 5th percentile. Hence, the averaging effect is contingent on the sizes of the weak and strong zones being much smaller than the size of the geotechnical system.

Since Pan et al. (2018a) used 200 realisations in their RFE studies, the lowest computed column capacity Q_{min} corresponds to the 0.5th percentile of the realisations. For the case of $CV_q = 0.3$, their results indicate that $\frac{Q_5}{Q_{min}} \sim 1.4$. This ratio corresponds approximately to the partial factor for unconfined compressive strength under the EC7's M2 approach, which uses material strength factoring. Hence, one may surmise that using Q_5 as the characteristic strength together with EC7's strength-factoring approach would lead to a probability of failure of the order of 0.5%. For higher CV_q, Q_{min} is likely to be lower and the probability of failure correspondingly higher. As a cautionary note, this Q_{min} estimate may be susceptible to increased errors as it is the extreme outlier; it should be regarded only as a very approximate estimate. Hence, while q_{u5} can still be used as a control strength parameter in quality assessment and control, in design, it may be prudent to set the characteristic strength at Q_5, which is about 88% and 70% of the 5th percentile unconfined compressive strength q_{u5}, under undrained and drained conditions, respectively.

6.7.1.1.2 Underground Treated Soil Slabs

An important difference between treated soil columns used as foundation elements and treated underground slab used for retaining wall support in deep excavations is the importance of positioning error. Treated soil columns used as foundation elements are often discrete; column-to-column contiguity is usually not a critical requirement. Although excessive positioning errors may still affect their load-carrying capacity, this is not a critical issue provided construction guidelines, such as those in Table 3.4, are adhered to. For underground treated soil slabs, column-to-column continuity is critical for the transmission of lateral forces. Excessive positioning error can lead to a loss of column-to-column contiguity, resulting in a decrease in slab stiffness and capacity. Hence, positioning error is a much more critical issue for underground treated soil slabs than for axially loaded treated soil columns.

As Table 3.4 shows, there are two components giving rise to positioning errors, namely errors in rig positioning and off-verticality. For deeply embedded underground slabs, off-verticality usually accounts for a larger proportion of positioning error at slab level. For instance, a 1% off-verticality will, at 20m depth, give rise to a positioning error of 200mm. This is significantly larger than typical construction limits on rig positioning error. Most construction guidelines for deep mixing and jet grouting allow for an off-verticality of between 1% and 2%, although 0.5% have been reported. This is similar to the situation for bored piling, which typically limits the off-verticality to 1.3% to 2%. Whatever is the adopted allowable limit, measures should be put in place to ensure that such limits are adhered to during construction of underground treated soil slabs.

In design, it may be prudent to assume a conservative level of positioning error since this will invariably affect the allowable position error during construction.

Table 3.4 indicates that a 2%, or approximately 1.15°, off-verticality may be assumed as an achievable level in most projects. For a treated soil column of 1m diameter, the normalised positioning error $\left(\xi/R\right)$ corresponding to 2% off-verticality for improvement depths of 10m and 20m are 0.4 and 0.8, respectively. This implies that, for 20m improvement depth, unless a lower level of off-verticality can be assumed in design, slab capacity may be significantly affected. For a 2m-diameter column, the normalised positioning error is halved that for 1m column. Hence, for deeply embedded treated soil slab, using larger soil columns can confer significant reduction in positioning error.

As Table 6.3 shows, Liu et al.'s (2015) and Pan et al.'s (2019) results indicate that, under undrained condition, with a CV of 0.4 on the unconfined compressive strength and 5% probability of failure, a positioning error $\left(\xi/R\right)$ of 0.5 gives an α-value (α_Q) on the slab capacity of about 1.0. This is equivalent to a strength reduction ratio $\left(Q_5/q_{uA}\right)$ of 0.53, which is significantly higher than the range of 0.35 and 0.42 for axially loaded columns under undrained condition. With a positioning error $\left(\xi/R\right)$ of 0.7, the strength reduction factor α rises to about 1.32 to 1.37, while the strength reduction ratio $\left(Q_5/q_{uA}\right)$ drops to between 0.35 to 0.38, this being similar to the undrained value for axially loaded soil columns. Hence, if the positioning error $\left(\xi/R\right)$ is well controlled, the 5th percentile slab capacity, which is also the equivalent strength of a homogeneous slab with 5% probability of failure, is likely to be higher than that of axially loaded column, in relation to the average UCS.

The strength ratio $\left(Q_5/q_{u5}\right)$ is also significantly higher than the corresponding values for axially loaded column. For a positioning error $\left(\xi/R\right)$ of 0.5, $\left(Q_5/q_{u5}\right)$ of treated soil slab is nearly twice that of axially loaded column. This is due to the much lower CV of slab capacity CV_Q, compared to that of column capacity. This can be attributed to the much larger dimensions of the underground slab, relative to the spatial variation SoF, which enables a much greater degree of averaging than in the case of an axially loaded column. Hence, the averaging effect is highly dependent on the dimensions of the treated soil domain, relative to the SoF. Larger dimension-to-SoF ratio enables better averaging, resulting in more consistent mass performance in the presence of spatial variation and smaller spread in the equivalent strength. Drained scenarios show a similar trend, although the differences are smaller.

For undrained condition and normalised positioning error $\left(\xi/R\right)$ of 0.7, the 5th percentile UCS, q_{u5}, is slightly higher than the 5th percentile equivalent strength, Q_5. For $\left(\xi/R\right)$ of 0.5, however, q_{u5} is less than Q_5 by more than 60%. This suggests that using 5th percentile unconfined compressive strength as an estimate for the 5th percentile equivalent strength may be overly conservative. On the other hand, the 10th percentile unconfined compressive strength, q_{u10}, overestimates Q_5 by about 9% for $\left(\xi/R\right)$ of 0.7 and underestimates it by about 26% for $\left(\xi/R\right)$ of 0.5. Thus, on balance, the 10th percentile unconfined compressive strength q_{u10} is probably a more representative estimate of Q_5.

For drained condition, the 5th percentile unconfined compressive strength q_{u5} is a good estimate for 5th percentile equivalent yield strength for $\left(\xi/R\right)$ of 0.5, but overestimates it by about 40% for $\left(\xi/R\right)$ of 0.7. This implies that, where drained condition prevails, the 5th percentile unconfined compressive strength may need to be factored down further for it to be representative of the 5th percentile equivalent yield strength. However, it should be noted that the equivalent yield strength merely represents a stress level at which the stiffness decreases significantly, rather than a maximum stress condition. Thus, it may be more appropriately regarded as a serviceability limit state rather than an ultimate limit state.

Using Liu et al.'s (2015) results, it can be shown that for an unconfined compressive strength CV (CV_{qu}) of 0.4, factoring Q_5 down by a strength mobilisation factor of 1.4, in line with EC7's strength factoring approach, will result in a probability of failure of 1×10^{-5} to 1×10^{-6}. Underground treated soil slabs used as temporary support for retaining walls in deep excavations are typically required to be functional between the start of excavation to the casting of the permanent base slab at formation level. This process typically does not require more than a few years. Hence, the probability of failure is roughly equivalent to the annual probability of failure for permanent structures. For dams and slope failures, Silva et al. (2008) noted that the annual probability of failure for dam and slope projects designed with a factor of safety of 1.5 according state-of-the-art engineering (Category I) and standard engineering practice (Category II) have probability of failure lying between 1×10^{-4} and 1×10^{-6}. Thus, a probability of failure of between 1×10^{-5} and 1×10^{-6} is equivalent to that for Categories I and II slope and dam projects. This is probably sufficiently low for most deep excavation works. Hence, for undrained condition, using q_{u10} as an estimate of Q_5, together with the EC7 strength mobilisation factors, may be considered to be adequate for deep excavation works.

6.7.1.1.3 Tunnel Improved Soil Surrounds

Table 6.5 summarises some published results on tunnel improved soil surrounds. All of the cases involved circular tunnel with circular improved soil surround with uniform thickness of 3m. Three different constitutive models were used in these studies, namely Mohr-Coulomb, Tresca and Cohesive Cam Clay. Tyagi et al. (2018), Pan and Lee (2019) and Pan et al. (2020) used the log-normal distribution to represent the distribution of unconfined compressive strength whereas Tyagi et al. (2020) used the β-distribution. The SoFs used were also different. Pan and Lee (2019) and Pan et al. (2020) assumed that the SoF is equal to the thickness of the soil surround, that is 3m. Tyagi et al. (2018) assumed that $SoF_x = SoF_y = 0.5$m, while Tyagi et al. (2020) modelled the columnar structure of the soil surround, with radial, circumferential and depth SoFs following those of Liu et al. (2015), which are much smaller than the thickness of the soil surround.

TABLE 6.5
Summary of some studies on tunnel improved soil surrounds.

References		Tyagi et al.'s (2018) 2D results using Mohr-Coulomb (SoF_x = SoF_y = 0.5m)		Pan and Lee's (2019) 2D results using Cohesive Cam Clay model		Tyagi et al.'s (2020) 3D results using Mohr-Coulomb: CV_q = 0.4		Tyagi et al.'s (2020) 3D results using Mohr-Coulomb: CV_q = 0.55		Pan et al.'s (2020) 2D results (for CV_q = 0.4) [Series R1]		
Drainage condition and names of parameters	**Parameters**	CV_q		CV_q		Normalised positioning error (ξ/R)		Normalised positioning error (ξ/R)		Improved soil model		
Undrained		**0.4**	**0.5**	**0.4**	**0.55**	**0.5**	**0.7**	**0.5**	**0.7**	**Cohesive Cam Clay**	**Tresca**	**Mohr-Coulomb**
Mean equivalent strength/mean UCS	Q_A/q_{uA}									0.64	0.66	0.63
CV of equivalent strength	CV_Q	0.20								0.16	0.11	0.13
5th percentile strength reduction ratio	Q_5/q_{uA}	0.40	0.35	0.51	0.40	0.40	0.36	0.40	0.33	0.50	0.54	0.48
5th percentile strength reduction factor (α)	α_{Q5}	1.50	1.30	1.22	1.10	1.51	1.59	1.10	1.21	1.26	1.15	1.31
Ratio of 5th percentile equivalent strength to 5th percentile UCS	Q_5/q_{u5}	0.81	0.85	1.04	1.05	0.92	0.85	0.92	0.78	1.01	1.10	0.97
		Drained: data only available for Pan et al. (2020)										
Mean equivalent strength/mean UCS	Q_A/q_{uA}									0.71	0.66	0.59
CV of equivalent strength	CV_Q									0.20	0.11	0.13
5th percentile strength reduction ratio	Q_5/q_{uA}									0.46	0.54	0.48
5th percentile strength reduction factor (α)	α_{Q5}									1.35	1.15	1.31
Ratio of 5th percentile equivalent strength to 5th percentile UCS	Q_5/q_{u5}									0.93	1.10	0.97

Entries without data are left blank.
Data sources: Pan and Lee (2019); Pan et al. (2020); Tyagi et al. (2018); Tyagi et al. (2020).

The 5th percentile strength reduction ratio Q_5/q_{uA} from these studies ranges from 0.33 to 0.54, overlapping with the corresponding values from axially loaded columns and underground slabs. This can be partly attributed to the differences in the soil models and the unconfined compressive strength distribution used. Pan et al. (2020) noted that the Mohr-Coulomb model generally gives more conservative results than Cohesive Cam Clay and Tresca models, due to its sloping shear yield surface, which reduces the shear yield strength as the effective stress decreases. This does not happen with the Cohesive Cam Clay, which has an arch yield surface, and the Tresca model, which has a flat yield surface.

Despite the differences in models and distributions, the results are generally consistent. In general, the 5th percentile equivalent strength Q_5 ranges from 0.78 to 1.1 times the 5th percentile unconfined compressive strength q_{u5}. Comparison of Tyagi et al.'s (2018) and Pan and Lee's (2019) two-dimensional results shows that the Mohr-Coulomb model gives more conservative values of Q_5 than the cohesive Cam Clay, as noted earlier. Furthermore, Tyagi et al.'s (2020) results show the important influence which the positioning error has on the results.

Pan and Lee's (2019) results give Q_5/q_{u5} values that are about 10% to 20% higher than Pan et al.'s (2018a) axial column results, whereas Tyagi et al.'s (2018) Q_5/q_{u5} values are similar to Pan et al.'s (2018a). This is because Tyagi et al. (2018) used the Mohr-Coulomb model whereas Pan et al. (2018a) and Pan and Lee (2019) used the cohesive Cam Clay model. Tyagi et al.'s (2020) results on improved soil surrounds overlap with Pan et al.'s (2018a) axially loaded column and Liu et al.'s (2015) underground slab results. For positioning error of 0.5, Tyagi et al.'s (2020) Q_5 values are slightly higher than Pan et al.'s (2018a) axial column Q_5 values. However, for positioning error of 0.7, Tyagi et al.'s (2020) Q_5 values are slightly lower than Pan et al.'s (2018a) axial column Q_5 values. The Mohr-Coulomb model used by Tyagi et al. (2020) explained their slightly lower Q_5/q_{u5} ratios than Pan and Lee's (2019). This indicates that, for the same soil model, the Q_5/q_{u5} ratio for improved soil surround is, in general, slightly higher than that for axially loaded column, by, say, about 10%. This can be readily attributed to the fact that the significant dimension for improved soil surround, that is its thickness, was assumed to be 3m in the cases studied. Since this is larger than the assumed column diameter of 1m, the corresponding averaging effect is also slightly stronger for the improved soil surround, compared to axially loaded column. In general, for improved soil surround of 3m or larger, q_{u5} may be considered a reasonably good estimate for Q_5. For thinner surrounds, q_{u5} may need to be factored down by 10% to 20%.

Pan et al. (2020) also estimated the probabilities of failure for the Mohr-Coulomb, Tresca and Cohesive Cam Clay, based on the log-normal distribution. Their results indicate that if Q_5 is factored down by 1.4, the resulting probability of failure may lie between 0.01% and 0.2%. This is higher than the probability of failure for underground slab. However, the duration over which the improved soil surround is required to provide temporary support is also likely to be shorter for tunnelling than for deep excavation. This is because tunnelling is a continuous operation, and excavation and installation of permanent lining typically proceed in tandem. Hence, the duration over which tunnel segment is supported only by the improved soil support is likely much shorter than that for deep excavations.

In summary, the foregoing discussion indicates that the 5th percentile equivalent strength Q_5 is suitable for use as a characteristic strength since it is not only complies with EC7's (BS EN 1997–1 2004) requirement that "the characteristic value should be derived such that the calculated probability of a worse value governing the occurrence of the limit state under consideration is not greater than 5%," but also allows an acceptably low probability of failure to be achieved if it is factored downwards using EC7's strength factoring guidelines. For axially loaded columns and thin improved soil surrounds with thickness less than 3m, the 5th percentile unconfined compressive strength q_{u5} may overestimate Q_5 by about 10% to 20%, owing to fluctuations induced by the SoF effect. Thus, some factoring down of q_{u5} may be prudent. For improved soil surround with thickness of 3m or larger, the averaging effect is slightly stronger and q_{u5} is a reasonably good estimate for Q_5. For underground slabs where the dimensions are much larger, the averaging effect is stronger and the 10th percentile unconfined compressive strength q_{u10} can be appropriate as an estimate of Q_5. Where good positioning control can be achieved, 20th percentile unconfined compressive strength q_{u20} may also be considered, albeit with some increase in the probability of failure. However, it should be noted that Liu et al.'s (2015) and Pan et al.'s (2019) studies did not consider the spatial variation arising from in situ water content variation, which probably has a much longer SoF than mixing quality. This may increase the fluctuation in the equivalent strength, that is CV_Q. Hence, some prudence should be exercised in raising the estimate of Q_5 from q_{u10} to q_{u20}.

In some cases, in situ variation in water content may have a high degree of determinacy. For instance, the Nicoll Highway Committee of Inquiry Report (Ministry of Manpower Singapore 2005) noted the possible presence of excess pore pressure in the marine clay underlying the sand fill in the reclaimed land. Jiang (2015) also noted that, in the MBFC project, the water content at the mid-depth of the marine clay appeared to be higher than that at the upper and lower boundaries. This is readily explained by the fact that the marine clay was still consolidating under the land fill. One way of dealing with such known or deterministic variation is to use different binder mix ratios.

6.7.2 Quality Control

In a typical construction project, the design is completed before the soil is treated. Hence, the properties of the treated soil are unknown at the design stage and the design would need to proceed based on an assumed value which has been shown to be safe and buildable, from past records. Because of this, it is important to ensure adequate quality control and record keeping, not just for control of the current project, but also as a basis for design in future projects.

An important item for quality control during construction, especially of underground treated soil slab, is the positioning error, which is influenced by rig positioning accuracy and off-verticality. Allowable limits of both these parameters should be clearly stipulated and records on them should be properly maintained. As mentioned earlier, for deep soil treatment, off-verticality often contributes more to positioning error than rig positioning error. One way to offset this in deep mixing is to install larger diameter columns.

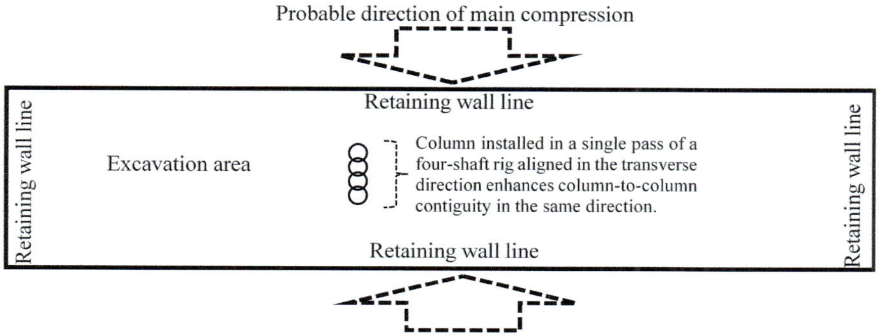

FIGURE 6.11 Illustrative example of aligning a multi-shaft mixing rig to enhance column-to-column contiguity. In this long, rectangular excavation, the main compression is likely to be in the direction transverse to the length since a long wall line is more flexible than a short wall line. Aligning the bank of shafts in the rig in the transverse enhances column-to-column contiguity in this direction since the columns formed in a single pass of the mixing rig are unlikely to have any relative positioning errors.

Liu et al. (2015) noted that, where multi-shaft rigs, such as the rig shown in Figure 1.15, are used in mixing, it may be advantageous to align the bank of shafts along the direction of main compression, if the latter can be identified. This ensures that the individual shafts in the rig are completely aligned, so that there is no relative positioning error between them. As Figure 6.11 illustrates, contiguity between individual columns formed by the different shafts in a single pass of the rig is automatically ensured, leading to enhanced degree of contiguity in the direction along the bank of shafts. This kind of scenarios may occur in long rectangular box type of excavation, such as that for some subway station boxes.

The importance of the coefficient of variation and scale of fluctuation has been highlighted in the foregoing discussion. The scale of fluctuation due to imperfect mixing is unlikely to vary significantly since it is constrained by the size of mixing column. The coefficient of variation, however, may vary depending upon the mixing methodology and equipment, soil conditions and workmanship. For this reason, from the quality assessment perspective, it is important to monitor the coefficient of variation. A change in the coefficient of variation may imply a change in methodology, soil conditions or workmanship. In a large project, where soil mixing is undertaken in several phases, the coefficient of variation may decrease with successive phases, indicating an improvement in mixing quality as the operators' understanding of the ground conditions improves.

Liu et al. (2017) showed that the minimum strength from a sample, that is sample minima, can fluctuate excessively as sample size increases. This indicates that there is some instability associated with the use of sample minima. The reason is that as the sample size increases, the chance of obtaining an outlying low strength sample can increase, rather than decrease. This is undesirable since, to be useful, a monitored parameter ought to converge towards a stable value as the sample size increases. It is also counterproductive since it discourages operators from getting more samples.

On the other hand, a percentile-type parameter, such as 5th or 10th percentile UCS, or a parameter derived by subtracting a multiple of the standard deviation from the mean, will not suffer from such fluctuations.

Liu et al.'s (2017) results also indicate that to achieve a stable value of the quality assessment parameter, a minimum of about 400 core samples is typically required. For small ground treatment projects, such a number may not economically feasible. On the other hand, for large projects, this number is usually well within the project budget. Whatever the case and where cost permits, operators should always be motivated to collect more, rather than less, samples for monitoring.

REFERENCES

Al-Naqshabandy, M. S. and Larsson, S. (2013). Effect of uncertainties of improved soil shear strength on the reliability of embankments. *Journal of Geotechnical and Geoenvironmental Engineering*, 139(4): 619–632.

ANCOLD (1998). *Guidelines on risk assessment*, Working Group on Risk Assessment, Australian National Committee on Large Dams, Sydney, New South Wales, Australia.

Ang, A. and Tang, W. (1975). *Probability concepts in engineering planning and design: Volume I—basic principles*, John Wiley and Sons, Inc., New York.

Ang, A. and Tang, W. (1984). *Probability concepts in engineering planning and design: Volume II—decision, risk, and reliability*, John Wiley and Sons, Inc., New York.

Baecher, G. B. and Christian, J. T. (2003). *Reliability and statistics in geotechnical engineering*, John Wiley & Sons, NJ.

BCJ (1997). *Design and quality control guideline for building foundation treatment: Deep/shallow soil mixing methods with cement admixtures*, Building Center of Japan (BCJ), Tokyo (in Japanese).

Broms, B. B. (1991). Stabilization of soil with lime columns. In *Foundation engineering handbook* (Ed. H. Y. Fang), 2nd Edition, Chapman & Hall.

Broms, B. B. (2004). Lime and lime/cement columns. In *Ground improvement* (Eds. M. P. Moseley and K. Kirsch), Vol. 2, Taylor & Francis, London: 252–330.

Bruce, M. E. C., Berg, R. R., Collin, J. G., Filz, G. M., Terashi, M. and Yang, D. S. (2013). *Federal highway administration design manual: Deep mixing for embankment and foundation support*. Department of Transportation Federal Highway Administration, McLean, VA, USA.

BS EN 1997–1 (2004). *Eurocode 7: Geotechnical design—Part 1: General rules*, European Committee for Standardization.

BS EN 12716 (2018). BS EN 12716: 2018. *Execution of special geotechnical work—jet grouting*. British Standards Institution.

BS-8006 (2010). BS 8006, 2010. *Code of practice for strengthened/reinforced soils and other fills*. British Standard Institution, UK.

Chen, J., Liu, Y. and Lee, F. H. (2016). A statistical model for the unconfined compressive strength of deep-mixed columns. *Géotechnique*, 66(5): 351–365.

Chin, K. G. (2006). *Constitutive behaviour of cement treated marine clay*. PhD thesis, National University of Singapore.

Ching, J., Hu, Y. G. and Phoon, K. K. (2016). On characterizing spatially variable soil shear strength using spatial average. *Probabilistic Engineering Mechanics*, 45: 31–43.

Croce, P. and Modoni, G. (2007). Design of jet-grouting cut-offs. *Proceedings of the Institution of Civil Engineers—Ground Improvement*, 11(1): 11–20. https://doi.org/10.1680/grim.2007.11.1.11.

Díaz-Francés, E. and Rubio, F. J. (2013). On the existence of a normal approximation to the distribution of the ratio of two independent normal random variables. *Statistical Papers*, 54: 309–323. https://doi.org/10.1007/s00362-012-0429-2.

EBGEO (2011). EBGEO, 2011. *Recommendations for design and analysis of earth structures using geosynthetic reinforcements* (Translation of the 2nd German Edition). German Geotech. Soc. ISBN: 978-3-433-02983-1, E-version ISBN: 978-3-433-60093-1.

El-Ramly, H., Morgenstern, N. and Cruden, D. (2002). Probabilistic slope stability analysis for practice. *Canadian Geotechnical Journal*, 39(24): 665–683.

EPA (2012). *U.S. EPA removal action Walton and Lonsbury site Attleboro, Massachusetts: Performance specification for soil mixing for ground improvement*. Rev. 2, 7 March 2012, https://semspub.epa.gov/work/01/524003.pdf.

Fenton, G. A. and Griffiths, D. V. (2007). Random field generation and the local average subdivision method. In *Probabilistic methods in geotechnical engineering* (Eds. Griffiths and Fenton), CISM International Centre for Mechanical Sciences (CISM), Vol. 491, Springer, Wien/New York. https://doi.org/10.1007/978-3-211-73366-0.

Filz, G. M. and Navin, M. P. (2006). Stability of column-supported embankments. *Final Contract Report, VTRC 06-CR13*, Virginia Transportation Research Council, Charlottesville, VA.

Griffiths, D. V. and Fenton, G. A. (Eds.) (2007). *Probabilistic methods in geotechnical engineering*, CISM International Centre for Mechanical Sciences (CISM) Vol. 491, Springer, Wien/New York. https://doi.org/10.1007/978-3-211-73366-0.

Grigoriu, M. (1993). On the spectral representation method in simulation. *Probabilistic Engineering Mechanics*, 8: 75–90.

Hasofer, A. A. and Lind, A. M. (1974). An exact and invariant second-moment code format. *Journal of Engineering Mechanics Division, ASCE*, 100: 111–121.

Hewlett, W. J. and Randolph, M. F. (1988). Analysis of piled embankments. *Ground Engineering*, 21(3): 12–18.

Jiang, Y. (2015). *Spatial variability in cement-treated marine clay—a field data study*. Master of Engineering dissertation, National University of Singapore.

Journel, A. B. and Huijbregts, C. J. (1978). *Mining geostatistics*, Academic, New York.

Karastanev, D., Kitazume, M., Miyajima, S. and Ikeda, T. (1997). Bearing capacity of shallow foundation on column type DMM improved ground. *Proc. 14th Int. Conf. Soil Mech. Found. Engrg.*, CRC Press, Hamburg: 1621–1624.

Kek, H. Y., Pan, Y., Ng, Y. C. H. and Lee, F. H. (2021). An approach for modelling spatial variability in permeability of cement-admixed soil. *Acta Geotech.,* 16: 4007–4026. https://doi.org/10.1007/s11440-021-01344-0.

Kitazume, M., Ikeda, T., Miyajima, S. and Karastanev, D. (1996). Bearing capacity of improved ground with column type DMM. *Proc. IS Tokyo '96/2nd Int. Conf. Ground Improvement Geosystems—Grouting and Deep Mixing*, Balkema, Rotterdom: Vol. 1: 503–508.

Kitazume, M. and Terashi, M. (2013). *The deep mixing method*, 1st Edition, CRC Press: 436pp.

Knabe, W., Przewlocki, J. and Rozynski, G. (1998). Spatial averages for linear elements for two-parameter random field. *Probabilistic Engineering Mechanics*, 13(3): 147–167.

Lee, F. H., Yong, K. Y., Quan, K. C. and Chee, K. T. (1998). Effect of corners in strutted excavations: Field monitoring and case histories. *Journal of Geotechnical and Geoenvironmental Engineering*, 124(4): 339–349.

Liedberg, N. S. D., Baker, S. and Smekal, A. (1996a). Samverkan mellan kalkcementpelare och lera (Interaction of lime/cement columns and clay). *Nordic Geotechnical Conference*, NGM-96, Reykjavik, Iceland: 43–48.

Liedberg, N. S. D., Baker, S., Smekal, A. and Ekström, J. (1996b). Samverkan mellan kalk-cementpelare och lera (Interaction between lime/cement columns and clay). *Swedish Railroad Authority, Technical Report 1996:2*, Chalmers Technological University, Report B1996:1, Department of Geotechnical Engineering: 107pp.

Little, D. N. and Nair, S. (2009). Recommended practice for stabilization of subgrade soils and base materials. *Contractor's Final Task Report for NCHRP Project 20-07*, National Cooperative Highway Research Program, Transportation Research Board: 57pp.

Liu, W. K., Belytschko, T. and Mani, A. (1986). Random field finite elements. *International Journal for Numerical Methods in Engineering*, 23(10): 1831–1845.

Liu, Y., He, L. Q., Jiang, Y. J., Sun, M. M., Chen, E. J. and Lee, F. H. (2019a). Effect of in situ water content variation on the spatial variation of strength of deep cement-mixed clay. *Géotechnique*, 69(5): 391–405.

Liu, Y., Jiang, Y. J., Xiao, H. and Lee, F. H. (2017). Determination of representative strength of deep cement-mixed clay from core strength data. *Géotechnique*, 67(4): 350–364.

Liu, Y., Lee, F. H., Quek, S. T. and Beer, M. (2014). Modified linear estimation method for generating multi-dimensional multi-variate Gaussian field in modelling material properties. *Probabilistic Engineering Mechanics*, 38: 42–53.

Liu, Y., Lee, F. H., Quek, S. T., Chen, E. J. and Yi, J.-T. (2015). Effect of spatial variation of strength and modulus on the lateral compression response of cement-admixed clay slab. *Géotechnique*, 65(10): 851–865.

Liu, Y., Li, J., Sun, S. and Yu, B. (2019b). Advances in Gaussian random field generation: A review. *Computational Geosciences*, 23: 1011–1047.

Ministry of Manpower Singapore (2005). *Report of the committee of inquiry into the incident at the MRT circle line worksite that led to the collapse of Nicoll Highway on 20 April 2004*. Ministry of Manpower, Singapore, 13 May 2005. https://www.nlb.gov.sg/main/book-detail?cmsuuid=7a71f7ff-272b-40df-98ef-cef463f3c663.

Nakagawa, S., Kamegaya, I., Kureha, K. and Yoshida, T. (1996). Case history and behavioural analyses of braced large scale open excavation in very soft reclaimed land in coastal area. In *Geotechnical aspects of underground construction in soft ground: Proceedings of the international symposium on geotechnical aspects of underground construction in soft ground* (Eds. R. J. Mair and R. N. Taylor), A. A. Balkema, Rotterdam, the Netherlands: 179–184.

Namikawa, T. and Koseki, J. (2013). Effects of spatial correlation on compression behavior of cement-treated column. *Journal of Geotechnical and Geoenvironmental Engineering (ASCE)*, 139(8): 1346–1359.

Namikawa, T. and Mihira, S. (2007). Elasto-plastic model for cement-treated sand. *International Journal for Numerical and Analytical Methods in Geomechanics*, 31: 71–107.

O' Rourke, T. D., McGinn, A. J., Dewsnap, J. and Stewart, H. E. (1998). Case history of an excavation stabilized by deep mixing methods. In *Design and construction of earth retaining systems* (Eds. R. J. Finno, Y. Hashash, C. L. Ho and B. P. Sweeney), GSP 83, American Society of Civil Engineers, Reston, VA, USA: 41–62.

Pan, Y. and Lee, F. H. (2019). Effect of spatial variability on undrained behaviour of tunnel with improved surroundings. *13th Int. Conf. Appl. Stat. & Prob. Civ. Engrg. (ICASP13)*, Seoul, South Korea.

Pan, Y., Liu, Y., Lee, F. H. and Phoon, K. K. (2019). Analysis of cement-treated soil slab for deep excavation support–a rational approach. *Géotechnique*, 69(10): 888–905.

Pan, Y., Liu, Y., Tyagi, A., Lee, F. H. and Li, D. Q. (2020). Model-independent strength-reduction factor for effect of spatial variability on tunnel with improved soil surrounds. *Géotechnique*, 71(5): 406–422. https://doi.org/10.1680/jgeot.19.P.056.

Pan, Y., Liu, Y., Xiao, H., Lee, F. H. and Phoon, K. K. (2018a). Effect of spatial variability on short-and long-term behaviour of axially-loaded cement-admixed marine clay column. *Computers & Geotechnics*, 94: 150–168.

Pan, Y., Shi, G., Liu, Y. and Lee, F. H. (2018b). Effect of spatial variability on performance of cement-treated soil slab during deep excavation. *Construction and Building Materials*, 188: 505–519.

Puppala, A. J., Madhyannapu, R. S. and Nazarian, S. (2008). *Special specification for deep soil mixing. Product 0-5179-P1*, The University of Texas at Arlington.

Rosenblueth, E. (1975). Point estimates for probability moments. *Proceedings of the National Academy of Sciences*, 72(10).

Shinozuka, M. and Deodatis, G. (1996). Simulation of multi-dimensional Gaussian stochastic fields by spectral representation. *Applied Mechanics Reviews*, 49(1): 29–53.

Shirlaw, J. N., Tan, T. S. and Wong, K. S. (2006). Deep excavations in Singapore marine clay. *5th Int. Conf. Geot. Aspects Underground Construction Soft Grd.* (Eds. Bakker, Bezuijen, Broere and Kwast), CRC Press, Amsterdam, Netherlands: 13–28.

Shirlaw, J. N., Wen, D., Nadarajah, P., Yoon, S. I. and Sugawara, S. (2000a). Construction issues related to jet grouted slabs at the base of excavations. *Int. Conf. Tunnels and Underground Structures*, Singapore, Singapore: 639–643.

Shirlaw, J. N., Wen, D., Nadarajah, P., Yoon, S. I. and Sugawara, S. (2000b). Design issues related to jet grouted slabs at the base of excavations. *Int. Conf. Tunnels and Underground Structures*, Singapore: 653–658.

Silva, F., Lambe, T. W. and Marr, W. A. (2008). Probability and risk of slope failure. *Journal of Geotechnical and Geoenvironmental Engineering*, 134(12): 1691–1699.

Suzuki, Y. (1982). Deep chemical mixing methods using cement as hardening agent. *Symposium on Soil and Rock Improvement Technology Including Geotextiles, Reinforced Earth, and Modern Piling Methods*, Bangkok, B-1-1-B-1-24.

Thyagaraj, T., Rao, S. M., Sai Suresh, P. and Salini, U. (2012). Laboratory studies on stabilization of an expansive soil by lime precipitation technique. *Journal of Materials in Civil Engineering*, 24(8): 1067–1075.

Topolnicki, M. (2016). General overview and advances in deep soil mixing. *Proc. XXIV Geotech. Conf. Torino—Design, Construction and Controls of Soil Improvement Systems*, Torino.

Tyagi, A., Liu, Y., Pan, Y. T. and Lee, F. H. (2020). Equivalent strength for tunnels in cement-admixed soil columns with spatial variability and positioning error. *Journal of Geotechnical and Geoenvironmental Engineering*, 146(10): 04020101.

Tyagi, A., Liu, Y., Pan, Y. T., Ridhwan, K. B. M. and Lee, F. H. (2018). Stability of tunnels in cement-admixed soft soils with spatial variability. *Journal of Geotechnical and Geoenvironmental Engineering*, 144(12): 06018012.

Tyagi, A. and Tamta, D. (2023). Equivalent design strengths for spatially variable cement-treated soil slope. *International Journal of Geomechanics*, 23(11): 06023017.

Tyagi, A., Xiao, H., Chin, K. G. and Lee, F. H. (2019). Model for predicting the unit weight of cement-treated soils. *Soils and Foundations*, 59: 1921–1932.

Tyagi, A., Zulkefli, M. F., Pan, Y, Goh, S. H. and Lee, F. H. (2017). Failure modes of tunnels with improved soil surrounds. *Journal of Geotechnical and Geoenvironmental Engineering*, 143(11). https://doi.org/10.1061/(ASCE)GT.1943-5606.0001788.

Xiao, H. (2009). *Yielding and failure of cement treated soil*. PhD Thesis. National University of Singapore, Singapore.

Xiao, H., Lee, F. H. and Chin, K. G. (2014). Yielding of cement-treated marine clay. *Soils and Foundations*, 54(3): 488–501.

Xiao, H., Lee, F. H. and Liu, Y. (2017). Bounding surface Cam-Clay model with cohesion for cement-admixed clay. *Int. Journal of Geomechanics*, 17(1): 04016026.

Yee, Y. W. and Tan, Y. C. (2015). Excavation support for TBM retrieval shaft using deep soil mixing technique, Kuala Lumpur. *Int. Conf. Exh. Tunn. Undergrd Space*, Petaling Jaya, Malaysia.

Yohannes, M. and Daramalingam, J. (2019). Specifying strength properties of soil mixing and jet grouting in Malaysia: A statistical perspective. *1st Malaysian Geot. Soc. (MGS) and Geot. Soc. Singapore (GeoSS) Conference*, Petaling Jaya.

Zienkiewicz, O. C. and Taylor, R. L. (2000). *The finite element methods*, Vol. 1, The Basis. Butterworth-Heinman, Oxford.

Index

Note: Page numbers in *italics* indicate a figure and page numbers in **bold** indicate a table on the corresponding page.

D

E

F

G

H

I

J

K

L

M

T

U

V

W

Y

Z

For Product Safety Concerns and Information please contact our EU representative GPSR@taylorandfrancis.com Taylor & Francis Verlag GmbH, Kaufingerstraße 24, 80331 München, Germany